Toxicology and Environmental Health

Toxicology and Environmental Health

Kaiden Higgins

STATES
ACADEMIC PRESS
www.statesacademicpress.com

States Academic Press,
109 South 5th Street,
Brooklyn, NY 11249, USA

Visit us on the World Wide Web at:
www.statesacademicpress.com

ISBN: 978-1-63989-529-8 (Hardback)

Cataloging-in-Publication Data

Toxicology and environmental health / Kaiden Higgins.
p. cm.
Includes bibliographical references and index.
ISBN 978-1-63989-529-8
1. Environmental toxicology. 2. Environmental health. 3. Toxicology. I. Higgins, Kaiden.
RA1226 .I58 2022
615.902--dc23

Table of Contents

Preface

This book aims to help a broader range of students by exploring a wide variety of significant topics related to this discipline. It will help students in achieving a higher level of understanding of the subject and excel in their respective fields. This book would not have been possible without the unwavered support of my senior professors who took out the time to provide me feedback and help me with the process. I would also like to thank my family for their patience and support.

The field of science that focuses on the harmful effects of diverse physical and biological agents on living organisms is called environmental toxicology. There are many sources that can lead to the presence of environmental toxicants. These include pesticides, organic and inorganic pollutants. Other sources comprises of heavy metals like arsenic lead, and cyanobacteria and cyanotoxins. Ecotoxicology is the branch of environmental toxicology which studies harmful effects of toxins on ecosystem and population. Aquatic toxicology is the multidisciplinary field that studies toxic effects of anthropogenic and natural activities on aquatic organisms and their ecosystem. This book attempts to understand the multiple branches that fall under the discipline of environmental toxicology and how such concepts have practical applications. The objective is to give a general view of the different areas of this field. This book, with its detailed analyzes and data, will prove immensely beneficial to professionals and students involved in this area at various levels.

A brief overview of the book contents is provided below:

Chapter - What is Environmental Toxicology?

Environmental toxicology deals with the study of the harmful effects of chemicals on the health of organisms and their habitat. Environmental toxins and contaminants, and bioaccumulation fall under its study. This is an introductory chapter which will briefly introduce about environmental toxicology.

Chapter - Sources of Environmental Toxicity

There are many sources of environmental toxicity that can lead to the presence of toxicants in our food, water and air. Some of them are polychlorinated biphenyl, lead, mercury, chromium, benzene, formaldehyde, radon, dichlorodiphenyltrichloroethane, etc. This chapter has been carefully written to provide an easy understanding of these sources of environmental toxicity.

Chapter - Ecotoxicology

Ecotoxicology is an integrated field of toxicology and ecology that studies the effects of toxic chemicals on biological organisms. A few of its concepts include microtox bioassay, dead zone, persistent organic pollutant, soil ecotoxicology, etc. This chapter discusses the subject of ecotoxicology in detail.

Chapter - Aquatic Toxicology

Aquatic toxicology is associated with the study of the harmful effects of chemicals on aquatic organisms and their ecosystems. Its major factors include ocean acidification, oil spilling in water bodies, aquatic contaminants and pollutants, etc. The topics elaborated in this chapter will help in gaining a better perspective of aquatic toxicology.

Chapter - Monitoring, Testing and Treatment

Various testing and treatment methods are used for the monitoring of environmental toxicity. A few of such methods are bioremediation, phytoremediation, environmental sampling, etc. All these monitoring, testing and treatment methods have been carefully analyzed in this chapter.

Kaiden Higgins

1

What is Environmental Toxicology?

Environmental toxicology deals with the study of the harmful effects of chemicals on the health of organisms and their habitat. Environmental toxins and contaminants, and bioaccumulation fall under its study. This is an introductory chapter which will briefly introduce about environmental toxicology.

Toxins

Toxins are substances that are poisonous to an organism. The term is sometimes restricted to poisons spontaneously produced by living organisms (biotoxins). Besides the poisons produced by such microorganisms as bacteria, dinoflagellates, and algae, there are toxins from fungi (mycotoxins), higher plants (phytotoxins), and animals (zootoxins). The name phytotoxin may also refer to a substance, regardless of origin, poisonous to plants.

Some biotoxins appear to be of little use to the organisms producing them but may play unknown roles in the organisms' metabolism, or they may be waste products. Many others, however, confer obvious advantages on their producers, such as depressing the growth of competitors or preventing predation. Many of the thousands of known phytotoxins are known to protect their producers against certain animals, especially insects. Similar defensive secretions are numerous among animals and may be either widely distributed among the tissues or concentrated in certain tissues, in which case there is often some sort of delivery system, such as spines or fangs.

Numerous animals (e.g., spiders and some snakes) have evolved poisons as aids in securing prey and often use the same poison for defense.

Many normally edible fishes and shellfishes become poisonous after feeding on toxic plants or algae. When a toxic fish or shellfish is eaten, the poison attacks the consumer's nervous system and causes a sometimes fatal condition called ciguatera.

Toxicant

A toxicant may be defined as an agent that causes an adverse effect or response in a biological system, seriously damaging its structure or function or producing death. The adverse effect or response may be defined in terms of a measurement that is outside the normal range for healthy organisms.

Introduction of Toxicants into the Ecosystem

A toxicant or foreign substance may be introduced deliberately or accidently into the ecosystem, impairing the quality of environment and making it unfavorable for organisms.

Toxicants find their way in the Ecosystem From

- Non-point sources such as agricultural runoff from land, contaminated ground water and bottom sediments, urban runoff, dredged sediment disposal, and atmospheric fall out.
- Point sources such as discharges (effluents) from manufacturing plants, hazardous waste disposal sites, and municipal waste treatment plants.

Survey and Classification of Toxicants

There are a number of chemical substances in the environment. Some of these are: Toxic and rest non-toxic. The toxicants are released from various sources into air, water and soil. They get into the human food chain from the environment. Once the toxicants enter our biological system, they interfere with the biochemical processes and may lead, in some, cases to fatal results.

According to a survey, there are four million known chemicals in the world today and another 30,000 new chemical compounds are added to this list every year. Among these, 60,000 – 70,000 chemicals are very commonly used. Apart from their benefit in increasing production, living standards and health, many of them are potentially toxic.

Survey of toxicants in various sections of environment may precisely by explained in the following points:

Toxicants in Air

Quite a large number of chemicals presumably pose serious problems of health when present in air. It is, therefore, necessary to exercise strict control during their manufacture and handling.

Environmental Protection Agency (EPA), listed 24 extremely hazardous substances in the atmosphere which are:

- Arsenic,
- Acrylonitrate,
- Asbestos,
- Berilium,
- Benzene,
- Cadmium,
- Chlorinated Solvents,
- Chromates,
- Chromates,
- Chlorofluorocarbons,
- Diethylstilbestrol,
- Dibromochloropropane,
- Ethylene Dibromide,
- Ethylene Oxide,
- Lead,
- Mercury,
- Ozone,
- Nitrosoamines,
- PCBs,
- PBBs,
- Sulphur Dioxide,
- Vinyl Chloride,
- Radiations,
- Toxic waste disposal emissions and leachates,

The common toxicants present in air are:

- Carbon Monoxide.
- Nitrogen Gases, viz., NO, NO_2, NH_3, and N_2O.
- Sulphur Gases, viz., SO_2 and H_2S.
- Hydrocarbons viz., CH_4.
- Photochemical Oxidants viz., O_3, and other photochemical products like benzopyrene, peroxybenzoyl nitrate (PB_ZN) and peroxyacetyl nitrate (PAN).
- Lead from automobile emission.

Toxicants in Water

A large number of toxicants — intentionally or otherwise — reach in the aquatic environment.

The principal sources of toxicants in aquatic environment are:

- Sewage and other wastes of domestic nature.
- Agricultural discharges.
- Industrial effluents.
- Wastes from thermal and nuclear power plants.

Toxicants in aquatic environment may, however, be grouped into two principal categories:

Toxic Trace Elements and Heavy Metals found in Natural and Waste Waters

Some of these are essential at low level, serving as nutrients for animal and plant life, but certainly deleterious at higher levels.

- Pesticides in water: The aquatic bodies contain a large number of pesticides, primarily from the drainage of agricultural field.

Toxicants Contaminating Food

A food additive may be defined as "a substance or mixture of substances other than a basic food, which is present in food as a result of any aspect of production, processing, storage or packaging". In other words, food additive is any substance not normally present in the food in question and which is added, either deliberately or incidentally, in order that some quality of the food may be improved.

Food additives may be of two types, namely, Intentional additives, for example, sodium benzoate, sorbic acid, lecithin, monosodium glutamate, saccharin etc., and Incidental additives, which are mixed with food during processing or packaging etc.

Food additives are used to provide better colour, flavour and texture and as a preservative. Several food additives have been found associated with the toxic hazards and health risks. Many of the additives like cyclamates, safrole and diethyl stilbestrol (DES) have been banned. Several others like saccharin, monosodium glutamate, nitrite, nitrate, etc. are under close investigation and are likely to be banned.

Others (Noise and Radiation Pollution)

Noise

Noise is an environmental pollutant, but it differs from air and water pollutants in that it disappears fast and thus does not remain in environment for long. Presence of unwanted sound in the atmosphere that produces adverse effect in human beings is termed as 'Noise Pollution'.

Vehicles do not just cause air pollution, they are also a major cause of noise pollution, exposure to which over an extended period of time can even lead to hearing loss.

Actually, sound, as such, is not considered as a pollutant. Many sounds are agreeable and pleasing to human ear. But the constant presence of unavoidable and irritating sound, for periods longer than the tolerance limit of human ear, is now considered as one of the serious air pollutants.

Noise can interfere with all our activities, such as work, rest, recreation, and sleep. It is not only a nuisance but also a threat to our health and, thus deserves the attention of all concerned.

Sources of noise are many. The main sources of man-made noise in developed urban areas are mechanized vehicles such as trucks, buses, cars, scooters, fire engines, tractors, generators etc. Industrial machinery, trains, aeroplanes, and accessory noise producers such as horns, sirens, loud-speakers etc. also contribute to noise pollution. Musical instruments, TV, radio, washing machines, vacuum cleaners, food mixers, air-conditioners, shouting, social gatherings, etc. are some other sources of noise pollution.

The main menace of noise, however, comes from transport. It mainly includes noise from road traffic, rail traffic, aircraft, in addition to vehicle horns. The number of road vehicles, particularly the diesel engine vehicles, has increased enormously in recent years.

Noise is transmitted as sound energy through any elastic medium like air, water or steel by the vibrations of the molecules in the medium. When there are no molecules, as in a vacuum, sound cannot travel. Sound travels faster in solids than in air. The speed of sound in air, at 20 °C, is nearly 344 m/s, while the speed of sound in wood is 3,962 m/s and in steel 5,029 m/s.

Sound is usually measured either by:

- Sound pressure.
- Sound intensity.

There are two important parameters for describing sound:

Sound Wave Frequency

As sound waves travel through air, the air undergoes periodic increase and decrease in density (compression and rarefaction). The number of compressions and rarefactions per unit time is known as the sound wave frequency. It determines the pitch of the sound. Its unit is Hertz (Hz) or cycles per second. The human auditory system is normally able to discriminate pitches in the range of frequencies between 16 and 20,000 Hz — called the audio-frequency range.

Sound Wave Amplitude or Magnitude

This is a measure of loudness. Since audible sounds may vary in intensity by many levels of magnitude, a logarithmic scale is used to measure sound intensity. This unit is decibel (dB), named after Alexander Graham Bell. The threshold for detection of sound is defined as 0 dB.

One dB is equal to the faintest sound a human ear can hear. Normal conversation generates about 50 to 60 dB, a city street has a sound level around 75 dB, and an auto horn at close range is about 90 dB. Sounds become painful and potentially damaging to auditory receptors at 130 dB. The most immediate and acute effect of noise pollution is impairment of hearing. Long exposure to loud noise can cause some hearing loss that may become permanent.

Generally, hearing loss is a result of neural damage involving injury to the hair cells. Two theories are offered to explain noise-induced injury. The first is that excessive hearing forces mechanically damage the hair cells. The second is that intense noise stimulation forces the hair cells into high metabolic activity that overdrives them to the point of metabolic failure and consequent cell death. Once destroyed, hair cells are not capable of regeneration. It has been established that a person exposed to a noise level of 90 dB for more than eight hours per day is liable to severe hearing problems after 20 years.

This is also true in case of traffic police personnel on duty and also for shopkeepers and residents near noise- polluted intersections. In addition to hearing damage, continuous noise can induce non-auditory physiological effects. Noise pollution can interfere with speech communication, sleep, acoustic privacy and cause annoyance — thus affecting human health, comfort and efficiency.

Noise pollution is also known to increase the rate of heart beat. It causes constriction of blood vessels, makes the skin pale, and muscles to contract leading to nervous breakdown and tension. It increases the digestive spasms and causes dilatation of the pupils of the eye.

Fluctuations in arterial blood pressure, decrease in heart output, impairment of night vision, and decrease in the rate of colour perception have also been reported. Noise also interferes in inter-personal communication. It causes headache, irritability (annoyance) and extreme emotional disturbances. It aggravates existing diseases by disturbing the peace of mind and sleep.

Effects of noise on the fetus are not fully known. Medical scientists have found that an unborn child moves and kicks about when there is loud noise. It also responds with a sudden increase in the heart beat as though it were disturbed or frightened. However, more research is required to establish a noise index consistent with human comfort and well- being.

Radiation Pollution

The atoms of certain elements are known to undergo spontaneous disintegration with the emission of atomic particles, i.e. radiation. Such elements or substances that emit radiations due to spontaneous nuclear disintegration are termed radioactive substances.

The well-known radioactive elements are:

Carbon^{14}, Thorium^{234}, Uranium^{238}, Phosphorus^{32}, Potassium^{40}, Cesium^{137}, Radon^{222}, Radium^{226} etc.

The radioactive rays and subatomic particles enter the environment mainly from two sources:

Natural Sources

The non-ionizing ultraviolet rays from the sun, cosmic rays and natural decomposition of radioactive substances emit radiation and sub-atomic particles.

Anthropogenic (Man-made) Radiation

- X-rays used in medical practice.
- Radiation therapy in treatment of cancer.
- Application of radio-isotopes in research work, medical practice and archeological dating.
- Atomic power plant.
- Nuclear generator.
- Radioactive fall-out resulting from nuclear weapons testing.

Contamination of the environment with the radioactive waves may be by:

- Accidental Spilling: Example—Chernobyl (Russia), Three Miles Island (U.S.A.) Nuclear disasters.
- Deliberate Action: Example — Atomic explosion as in Hiroshima, Nagasaki (Japan) and explosives in wars.
- Radioactive Substances: Example — Cesium, Uranium, Zirconium, Tritium, Strontium, Radium, Plutonium etc.
- Radio-Isotopes: Example — Isotopes of Uranium^{238}, Carbon^{14}, Cobalt^{60}, Manganese^{54} and Strontium^{90}.
- Ignorance: Example — Disposal of nuclear wastes indiscriminately and cutting of forests.

Types of Radiation

Radiations are chiefly of two types:

- Non-ionizing UV.
- Ionizing (X-rays, alpha, beta, neutrons and gamma radiations).

Ionizing radiations are of two types:

Electromagnetic Radiation

It is in the form of short wave, high energy radiation similar to X-rays. Example, gamma radiation.

Corpuscular or Particulate Radiation

It consists of subatomic particles which transfer their energy to whatever object they strike. Example, alpha and beta radiations.

Radiation Effects

Non-ionizing Radiation

The radiations which do not possess ionizing effect are called non-ionizing radiation.

UV Rays

UV rays emitted by sun may penetrate living tissues up to 0.1 mm thickness, depending upon the length of exposure and are specifically absorbed by nucleic acids and generate pyrimidine dimers. These suppress separation of DNA strand during replication. These may produce mutation.

Ionizing Radiations

Very high-energy radiations that are able to remove electrons from atoms and attach them to other atoms, thereby producing +ve and -ve ion pairs, are known as ionizing radiations. Isotopes of elements that emit ionizing radiations are called radionuclides or radioisotopes.

X-Rays

The X-rays possess high speed almost equal to the speed of light – and have ionizing power but do not penetrate deep in the living tissues.

Gamma Rays

The gamma rays possess enormous penetrating power and travel with a speed that equals speed of light. It may penetrate a lead sheet of about 15 cm (6 inch) thickness. It also damages living tissues.

Classification of Toxicants on the basis of Organ/System Affected

From the human health view-point and the organ/system affected, the toxicants may be of the following types:

Environmental Carcinogens

In various countries, cancer is one of the leading cause of death. It is supposed to be induced by a variety of agents and also certain life styles, e.g., smoking cigarettes, drinking alcohol, chewing tobacco etc. However, toxicants in the community and at work places, besides certain therapeutic agents, also play certain role in inducing cancer.

Attempts at reducing cancer incidence are, therefore, aimed at minimizing these hazardous life styles along with identifying these chemical carcinogens, in the hope that human exposure to them can be eliminated or minimized. Chemical carcinogens may thus be defined as any chemical which initiates carcinogenicity.

Cardiotoxicants

Heart may not be a common target organ, yet it is damaged by a variety of chemical or drugs referred to as cardiotoxicants. They act either directly on the myocardium or indirectly through the nervous system or blood vessels.

Immunotoxicants

Actually, the immune system serves an important role in the body's defense against infection by microbes viz., viruses, bacteria, fungi and unicellular and multicellular parasites as well as against neoplastic cells.

It has now become an established fact that a variety of toxicants (chemicals) can impair the function of the system. A variety of chemicals may induce hypersensitivity reactions. The site of such reactions are mainly confined to the respiratory tract and the skin. Dean et al reported that a variety of substances may adversely affect the immune system.

Teratogens

Chemical agents causing teratogenesis are called teratogens. Teratogenesis is the formation of congenital defects.

Occupationally Inhaled Toxicants

Following industrialization, since the Industrial Revolution begun in 1970, the human respiratory system is increasingly exposed to airborne toxicants. Airborne substances exist in the form of gases, vapours, liquid droplets and solid particulate matters of different sizes. The uptake and effects of inhaled toxicants, therefore, depend not only on their toxic nature but also on their physical characteristics.

Hepatotoxicants

Some of the hepatic (liver) injuries are produced by hepatotoxicants like CCl_4; Halothane; α-naphthylisocyanate (ANIT); Chloroform, Ethanol; Bromobenzene; Bromsulphathalein (BSP); Aflatoxins and Phosphorus. These hepatotoxicants induce several types of injuries like steatosis, necrosis, cirrhosis, neoplasia etc. These injuries are relatively easily produced in experimental animals.

Mutagenic Agents

Mutagenesis can occur as a result of interaction between mutagenic agents (like As, Cd, Co, Hg, Ni, Pb and Pt) and the genetic material, i.e., DNA of organisms. While spontaneous mutation and natural selection are the major means of evolution, a number of toxicants in recent decades have been reported to be mutagenic.

Environmental Toxicology

Environmental Toxicology is a field of study in the environmental sciences that is concerned with the assessment of toxic substances in the environment. Although it is based on toxicology, environmental toxicology draws heavily on principles and techniques from other fields, including biochemistry, cell biology, developmental biology, and genetics. Among its primary interests are the assessment of toxic substances in the environment, the monitoring of environments for the presence of toxic substances, the effects of toxins on biotic and abiotic components of ecosystems, and the metabolism and biological and environmental fate of toxins.

Environmental toxicology is a relatively young field, with its origins in the mid-20th century. The modern science of toxicology, on the other hand, was born in the early 19th century, and by the later decades of that century, some scientists had begun to consider the effects of toxic substances that had been released into the environment. But awareness of environmental pollutants did not increase markedly until the publication of American biologist Rachel Carson's Silent Spring in 1962. Despite strong opposition from the chemical industry, which felt that Carson's work unfairly attacked their products, Carson highlighted the environmental side effects from the use of pesticides such as DDT. The book suggested that pollutants used in one area could quickly affect neighbouring areas and that the destruction of a particular part of the food chain upsets the balance of nature, leading to the destruction of an ecosystem. In 1969 scientist René Truhart coined the term ecotoxicology to describe the study of the toxic effects of pollutants on the biological components of ecosystems. Although narrower in scope, ecotoxicology played an important role in the development of environmental toxicology.

In the 1970s scientists concerned with toxins in the environment increasingly focused their research on the impacts of agents of biological warfare (e.g., Agent Orange), on industrial pollution, and on discharge from mines. In the next decade the relevance of environmental toxicology to modern society was realized with the Bhopal disaster, in which 45 tons of poisonous methyl isocyanate gas escaped from an insecticide plant, and the Chernobyl accident, in which massive amounts of radioactive material were released into the atmosphere following the explosion of a nuclear reactor. The Bhopal disaster killed between 15,000 and 20,000 people, with many thousands more suffering from acute and chronic conditions. The Chernobyl accident ultimately was responsible for an estimated several thousand deaths from radiation sickness and cancer and the contamination of millions of acres of land across Belarus, Russia, and Ukraine. In the late 20th and early 21st centuries, the field of environmental toxicology expanded. Among its major concerns were oil spills, the dumping of medical and nuclear waste, air and water pollution, and the impact of substances such as synthetic hormones that were regularly released into environmental reservoirs.

Assessment and Monitoring of Toxic Substances

Toxins affect the environment and organisms in a variety of ways, from having little negative impact on certain abiotic factors or resistant organisms to killing animals and destroying major components of ecosystems. The extent of damage depends on the type and structure of the toxic substance; the age, the size, and the species of the organism; and the temperature and the physical and chemical characteristics of the environment (whether terrestrial or aquatic). Knowledge of how these factors interact is critical to understanding how best to prevent or reduce exposure or remove a toxin from the environment (environmental remediation).

The assessment of toxicity at the levels of whole organism, cell, and gene is one way by which researchers are able to determine how much of a toxin an organism can be exposed to before adverse effects set in. Different assays are used for toxicity assessment, including acute and subacute toxicity assays, sediment toxicity assays, and genotoxicity assays. The determination of safe exposure levels in animals plays a key role in the development of regulations that dictate how toxic substances are to be handled and disposed of. There are also methods by which scientists are able to estimate the quantity of a given toxic substance in the environment.

The identification of ways to monitor for chemicals in the environment is an important aspect of environmental toxicology. Monitoring typically is based on the detection of sensitive biochemical markers (e.g., proteins), the levels of which change in the presence of a given toxin, or on changes in individual "indicator" species, the well-being of which serves as a measure of environmental conditions and the health of other species.

Environmental Toxins

Environmental toxins are cancer-causing chemicals and endocrine disruptors, both human-made and naturally occurring, that can harm our health by disrupting sensitive biological systems.

Environmental toxins include naturally occurring compounds such as:

- Lead.
- Mercury.
- Radon.
- Formaldehyde.
- Benzene.
- Cadmium.

They also include human-made chemicals like:

- BPA.
- Phthalates.
- Pesticides.

In toxic doses, all of these compounds can negatively affect human health. Many of them are known to:

- Cause cancer (radon, formaldehyde, benzene).
- Act as endocrine disruptors (bpa, pesticides, phthalates).
- Cause organ failure or developmental problems (lead, mercury, cadmium).

Lead toxicity is a well-known example. People are generally aware of potential sources of lead, such as old paint and old pipes. Cadmium toxicity was first realized in the 50s and 60s, and policies now limit industrial exposure. Mercury is also a well-known toxin.

Endocrine Disruptors

Endocrine disruptors include a wide range of substances, both natural and human-made, that may interfere with the body's endocrine (hormone and cell signaling) system and produce adverse developmental, reproductive, neurological, and immune effects.

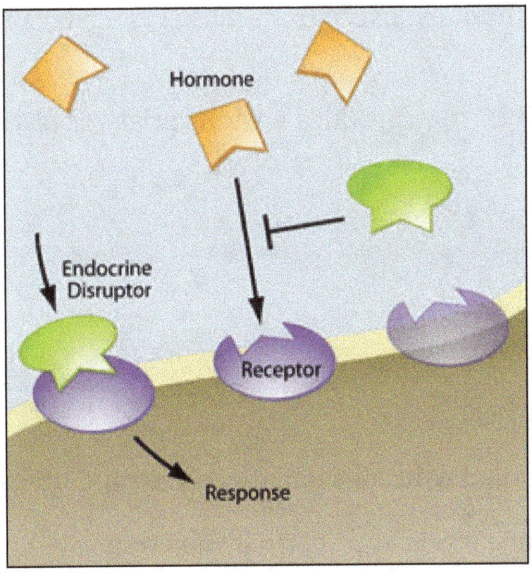

Action of endocrine disruptors.

Endocrine disruptors usually mimic estrogen and are found in many everyday products we use, including:

- Some plastic bottles and containers.
- Food can liners.
- Detergents.
- Flame retardants.
- Toys.

- Cosmetics.
- Pesticides.

In particular, the industrially produced compounds bisphenol A (BPA), pesticides, and phthalates are among the most potentially dangerous.

Research shows that endocrine disruptors may pose the greatest risk during prenatal and early postnatal development when organ and neural systems are forming. Pregnant or breast-feeding women, or women planning on becoming pregnant, should be the most cautious.

BPA: Bisphenol A

Much of the concern about endocrine disruptors has focused on BPA, a compound that is widely used in the manufacture of polycarbonate plastics and epoxy resins that are used in food and drink packaging, water and baby bottles, metal can linings, bottle tops, and water supply pipes. In addition, BPA can also be found in thermal paper receipts, though the amount of exposure from these particular products is thought to be minimal.

Low-dose exposure to BPA may produce a wide variety of physiological problems, including:

- Obesity.
- Infertility.
- Aggressive behavior.
- Early onset of puberty.
- Hormone-dependent cancers such as prostate and breast cancer.
- Lower testosterone levels and sperm production.

BPA exposure occurs when the chemical leaches out from the product into food and water, especially when plastic containers are washed, heated or stressed. The highest estimated daily intakes of BPA occur in infants and children.

In fact 93% of children 6 years of age and older have detectable levels of BPA in their urine, and a 2011 study found that 96% of American women also have detectable levels. In September 2010, Canada became the first country to declare BPA a toxic substance; the European Union and Canada now ban BPA in baby bottles.

The good news is that BPA exits the body quickly. A 2011 study found that when participants ate their usual diets, followed by three days of consuming foods that were not canned or packaged, BPA levels in their urine decreased by 66%.

Reduce Exposure to BPA

- Minimize use of plastic containers with the #7 or #3 on the bottom.
- Don't microwave plastic food containers, and don't wash them in the dishwasher or with harsh detergents.
- Reduce use of canned foods and eat mostly fresh or frozen foods.
- When possible opt for glass, porcelain or stainless steel cups, containers, water bottles and travel mugs.
- Use baby bottles that are BPA free (or better yet use glass bottles) and look for toys labeled BPA free.

Pesticides

Pesticides are any substance used to kill, repel, or control certain forms of plant or animal life that are deemed pests. This includes herbicides, insecticides, fungicides, disinfectants, and compounds used to control rodents. In the US, over 4.5 billion pounds of pesticides are used each year.

Most conventional food production uses pesticides, so people are exposed to low levels of pesticide residues through their diets. While the health effects of pesticide residues are not entirely clear, research from the National Institute of Health showed that farmers who use agricultural insecticides experience an increase in headaches, fatigue, insomnia, dizziness, hand tremors, and other neurological symptoms, while licensed pesticide applicators have a 20-200% increased risk of developing diabetes.

Other data found that individuals reporting regular exposure to pesticides had a 70% higher incidence of Parkinson's disease than those reporting no exposure.

It also appears that children are particularly susceptible to adverse effects from exposure to pesticides, specifically neuro-developmental problems. This is probably because children eat more food relative to their size. They also play in the dirt and spend time on the ground, where pesticides may linger.

Reduce Exposure to Pesticides

- Wash and scrub all fruits and vegetables, organic or conventional.
- If possible purchase mostly organic fruits and vegetables, particularly the ones consistently found to have the highest pesticide residues – apples, strawberries, celery, peaches and spinach.
- Grow your own.

Phthalates

Phthalates are chemicals used to soften plastics. They are found in a wide variety of products, including bottles, shampoo, cosmetics, lotions, nail polish, and deodorant. At one time most flexible plastics contained high levels of phthalates. Fortunately, they are being phased out in the US and Europe due to emerging recognition of their risks.

The National Institute of Environmental Health Sciences (NIEHS), part of the National Institute of Health, has found that pre-natal exposure to phthalates is associated with adverse genital development and can significantly reduce masculine behavior in boys. Women with high exposure to phthalates while pregnant report significantly more disruptive behavior in their children, while other research by NIEHS has found phthalate exposure can lead to thyroid dysfunction in adults.

Fortunately, as with BPA, if exposure is decreased, phthalates quickly exit the body. The same study that found a large decrease in BPA levels a mere three days after participants stopped eating canned and packaged foods also found that phthalate levels in the urine decreased by 53-56% during the same time period.

Reduce Phthalate Exposure

- Minimize use of plastics with the recycling code #3.
- Use PVC-free containers: Buy plastic wrap and bags made from polyethylene and use glass containers. If you do use plastic containers, do not heat or microwave them.
- Choose phthalate-free toys: Many large toymakers have pledged to stop using phthalates, but be sure to look for toys made from polypropylene or polyethylene.
- Purchase phthalate-free beauty products: Avoid nail polish, perfumes, colognes, and other scented products that list phthalates as an ingredient. Many scented products simply list "fragrance" as an ingredient, which often incorporates a number of different chemicals including phthalates.

Carcinogens

Hundreds of chemicals are capable of inducing cancer in humans or animals after prolonged or excessive exposure. Chemically-induced cancer generally develops many years after exposure to a toxic agent. For example, mesothelioma (a form of lung cancer) may take 30 years to emerge after asbestos exposure.

In 2010, the U.S. President's Cancer Panel Report declared:

> "The true burden of environmentally induced cancers has been grossly underestimated, this group of carcinogens has not been addressed adequately by the National Cancer Program. The American people – even before they are

born – are bombarded continually with myriad combinations of these dangerous exposures."

According to the report there are about 80,000 chemicals in commercial use in the United States, but only about 2% of those have been assessed for their safety.

The Cancer Panel report singles out radon, formaldehyde, and benzene as major environmental toxins that are causing cancer.

Radon

Radon is a colorless, odorless radioactive gas. It comes from the natural decay of uranium or thorium found in nearly all soils and it typically moves up through the ground and into the home through cracks in floors, walls, and foundations.

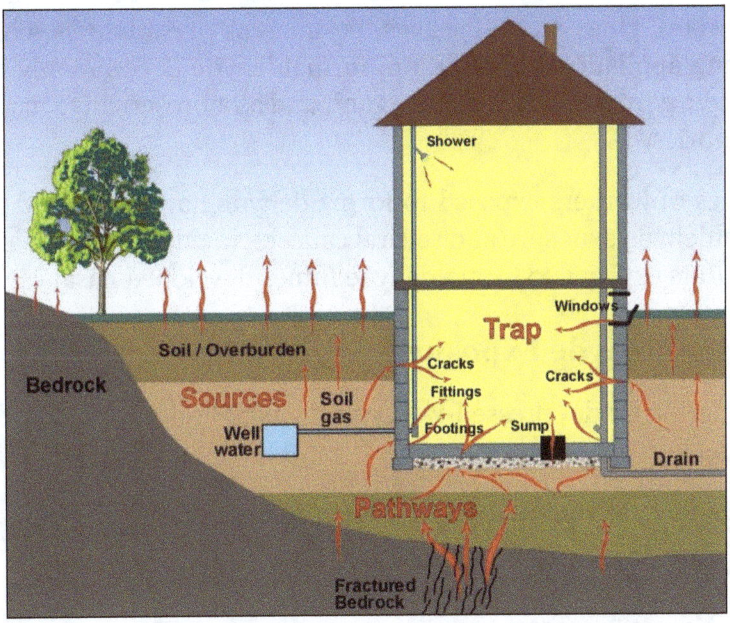

Radon movement from ground to home.

It can also be released from building materials or from well water. Radon breaks down quickly, giving off radioactive particles. Long-term exposure to these particles can lead to lung cancer.

The U.S. Environmental Protection Agency estimates that radon causes about 21,000 lung cancer deaths in the United States each year, with 1 in 20 US homes having elevated levels. Radon exposure is the second leading cause of lung cancer after smoking and the leading cause among non-smokers.

Many radon-related lung cancer deaths can be prevented by testing for radon and taking the necessary steps to lower radon levels in homes that have elevated radon. This process is known as radon mitigation.

Reduce Radon Exposure

- Get your home air checked. It is simple and inexpensive.
- If you use a well, check your water also.

Formaldehyde

Formaldehyde is a colourless, flammable, strong-smelling chemical that is used in building materials and in the manufacture of many household products. It also occurs naturally in the environment and is produced in small amounts by most living organisms as part of normal metabolic processes. Several government agencies have classified formaldehyde as a known human carcinogen.

Formaldehyde sources in the home include pressed-wood products such as particleboard and plywood, glues and adhesives, permanent press fabrics, cigarette smoke, and fuel-burning appliances. In addition, formaldehyde is commonly used as an industrial fungicide, germicide, and disinfectant, and as a preservative in mortuaries and medical laboratories.

Research studies of workers exposed to formaldehyde have suggested an association between formaldehyde exposure and several cancers, including nasopharyngeal cancer and leukemia. Rats exposed to formaldehyde fumes developed nasal cancer.

Reduce Formaldehyde Exposure

- Use "exterior-grade" pressed-wood products to limit formaldehyde exposure in the home.
- Ensure adequate ventilation and moderate temperatures.
- Reduce humidity levels with air conditioners and dehumidifiers.
- Go natural and grow plants in your home.

Benzene

Benzene is a colourless liquid that evaporates quickly. It is naturally found in crude oil and is a basic petrochemical. Unfortunately, it is also a known human carcinogen.

Benzene is found in tobacco smoke, gasoline (and therefore car exhaust), pesticides, synthetic fibres, plastics, inks, oils, and detergents. Benzene has also been found in dryer emissions from scented laundry detergent and dryer sheets, and in soft drinks, although these have since been reformulated to exclude it.

About 50% of the benzene exposure in the US results from smoking tobacco or from second-hand smoke.

Substantial amounts of data link benzene to aplastic anemia, bone marrow abnormalities, and leukemia — particularly acute myeloid leukemia (AML) and acute non-lymphocytic leukemia (ANLL).

Reduce Benzene Exposure

- Don't smoke and try to avoid second hand smoke.
- Ensure adequate ventilation in your home.
- Use non-scented laundry detergents.
- Keep plants in the home.

It is important to remember that the poison is in the dose. Problems usually result from prolonged or excessive exposure; the occasional use of a plastic cup probably won't hurt you.

While it is impossible to completely eliminate exposure (and it might drive you crazy to try!), a few simple steps will go a long way towards protecting you and your family:

- Decrease use of plastic – Transition to glass, stainless steel and porcelain containers, glasses and mugs.
- Wash all produce, and if possible purchase organic options from the Dirty Dozen.
- Use fewer products with the term "fragrance."
- Get your home air and water checked for radon.
- Don't smoke.
- Keep plenty of plants in the home.

There's no need to freak out over occasional exposure to environmental toxins. Just look for simple ways to reduce your everyday exposure. Make changes slowly, one at a time, in a manageable way, and you will decrease your risk with minimal stress.

Environmental Contaminants

The term environmental contaminants refers to harmful chemicals present in soil, air and water. These compounds may come directly from human sources such as industrial manufacturing, agricultural run-off and wastewater discharge, or they may originate from natural sources, such as the taste and odor-causing chemicals in water generated by algaeand bacteria blooms.

NDMA (N-nitrosodimethylamine)

The nitrosamine group, a group that includes Nnitrosodimethylamine (NDMA), has attracted increased attention from the water treatment community as potent and potentially wide-ranging water contaminants. While their presence in municipal wastewater and indirect potable reuse is relatively well understood, understanding of their presence in drinking water is limited.

Taste and Odor Compounds

The primary sources of taste and odor problems in drinking water are algae and bacteria. However, other anthropogenic sources such as wastewater discharges and chemical spills also act as sources of chemicals that cause off tastes and odors. Such chemicals can affect both groundwater and surface water.

Health Effects of Toxic Chemicals

There are many materials used in the workplace that can be hazardous. However, in order for them to affect your health, they must come into contact with the body or be absorbed into the body. When assessing the potential health effects from working with a particular material, it is necessary to understand the difference between "toxicity" and "hazard".

- Toxicity is the ability of a substance to produce an unwanted effect when the chemical has reached a sufficient concentration at a certain site in the body. The more toxic a material is, the smaller the amount of it is necessary to be absorbed before harmful effects are caused. The lower the toxicity, the greater the quantity of it is necessary to be absorbed. Generally, the toxicity of a chemical is determined through experiments on animals (quite often rats), and measured in terms of the amount of material necessary to cause death in 50% of the test animals. These values are called LD50 (lethal dose) or LC50 (lethal concentration), and are usually given in weight of material per kg of body weight or airborne concentration of material per set time period respectively.
- Hazard is the probability that this concentration in the body will occur. Toxicity is an inherent property of the material. A material may be very toxic, but not hazardous, if it is handled properly and is not absorbed into the body. On the other hand, a material may have a very low toxicity, but be very hazardous.

Example:

- An open container of an acid is much more hazardous than a closed container of the same material.

- Two liquids may possess the same degree of toxicity but present different degrees of hazard.
- One material may be non-irritating to the eyes and nose and odourless. The other may be irritating to the eyes or respiratory system and possess a pungent odour.
- The latter material, because of its warning properties presents a lesser degree of hazard.

In order for toxicants to affect the human system either they must cause damage to external tissues, such as the skin or eyes, or they must be able to enter the body by some mechanism.

Routes of Entry

There are three primary routes of entry into the body: Ingestion, Skin or Eye Absorption, Inhalation and Injection.

- Ingestion: This means taking a material into the body by mouth (swallowing). Ingestion of toxic materials may occur as a result of eating in a contaminated work area.
- Absorption: Substances that contact the eye and the skin may be either absorbed into the body or cause local effects. For the majority of organic compounds, the contribution from skin absorption to the total exposure should not be neglected.
- Inhalation: This means taking a material into the body by breathing it in. In the lungs, very tiny blood vessels are in constant contact with the air we breathe in. As a result, airborne contaminants can be easily absorbed through this tissue. In the occupational environment, this is generally the most important route of entry.
- Injection: This means taking in biological or chemical substances into the body when the skin is accidentally punctured with a sharp object (e.g. contaminated needle).

Health Effects: Chronic vs. Acute

Once a toxic substance has contacted the body it may have either acute (immediate) or chronic (long term) effects.

Example: Spilling acid on your hand will cause immediate harm, i.e. a burn to the skin.

Exposure to asbestos or tobacco smoke may result in lung cancer after as much as twenty years (this is a long term effect).

Exposure: Chronic vs. Acute

Exposure can be classified as chronic or acute. In chronic exposures, the dose is delivered at some frequency (daily or weekly usually) over a period of time. In acute exposures, the dose is delivered in a single event and absorption is rapid. Usually, a chronic exposure occurs at low concentration and acute exposure at high concentration.

Some materials may only cause harm if given acutely, not having any effect in the long term. Other materials may not exhibit an effect in the short term, but may cause problems after prolonged exposure.

Physiological Classification of Materials

This classification identifies toxic materials on the basis of biologic action:

- Irritants: Refers to some sort of aggravation of whatever tissue the material comes in contact with. e.g. ammonia, nitrogen dioxide.
- Asphyxiants: Exert their effects through a depletion of oxygen to the tissues e.g. simple asphyxiants – carbon dioxide, nitrogen, methane, hydrogen chemical asphyxiants – carbon monoxide, hydrogen cyanide, hydrogen sulphide.
- Narcotics or Anaesthetics: The main toxic action is the depressant effect upon the Central Nervous System. e.g. – many organics, chloroform, xylene.
- Systemic Poisons: The main toxic action includes the production of internal damage e.g. Hepatotoxic agents – toxic effects produce liver damage. eg. carbon tetrachloride. e.g. Nephrotoxic agents – toxic effects produce kidney damage eg. some halogenated hydrocarbons.
- Carcinogens: Agents/compounds that will induce cancer in humans. e.g. benzene, arsenic, inorganic salts of chromium, nickel, beryllium.
- Mutagens: Agents that affect the cells of the exposed people in such a way that it may cause cancer in the exposed individual or an undesirable mutation to occur in some later generation. e.g. radiation, variety of chemical agents that alter the genetic message.
- Teratogens: Agents or compounds that a pregnant woman takes into her body that generate defects in the fetus e.g. Thalidomide, possibly steroids.
- Sensitizers: Agents that may cause allergic or allergic-like responses to occur.

After an initial exposure to a substance an individual may become sensitized to that substance. Subsequent exposures to the same substance, often at a much lower concentration than before, produces an allergic response. This response may be a skin rash (dermatitis) or an asthmatic-like attack, depending on the route of exposure. e.g.

cutting oils, isocyanates in polyurethane foam operations and paint spraying operations, some laboratory solvents.

Bioaccumulation

Bioaccumulation is defined as the net accumulation of a contaminant in or on an organism from all sources including water, air, and diet.

Bioaccumulation is defined as the increase of contaminant concentrations in aquatic organisms following uptake from the ambient environmental medium. Different sources of exposure contribute to contaminant bioaccumulation. Toxicokinetic parameters are critical in the bioaccumulation processes, including the dissolved uptake rate constant, dietary assimilation efficiency, and efflux. Much study has been conducted on the quantification of these biokinetic parameters in diverse marine organisms. A well-developed biokinetic model then incorporates these biokinetic processes as well as contaminant geochemistry and animal physiology to predict bioaccumulation of contaminants in aquatic organisms. There are many applications of the biokinetic model in aquatic ecotoxicology, especially in understanding the change of contaminant concentration, which forms a fundamental basis in aquatic toxicology. Although biomonitoring programs have been established for decades, there are several caveats underlying the application of biomonitors given the multiple ecosystem stresses currently faced by coastal and estuarine ecosystems.

Bioaccumulation can be considered as a particular type of biosorption in which metals are incorporated inside living biomass. This incorporation relies mostly on active uptake, but passive uptake can also be involved. Metals are incorporated within biomass as a mean of protection against metal toxicity. Metals can be immobilized through complexation with intracellular proteins such as sulfur-containing peptides (metallothionein and phytochelatins). Alternatively, in eukaryotic organisms, metals can be stored in intracellular organelles. As a result, bioaccumulation relies strongly on an organism affinity toward a given metal, as well as on the physicochemical conditions influencing metal toxicity (Eh, pH). Metal accumulation into fruiting bodies of mushrooms is a well-known aspect of the fungal storage capabilities for various compounds. For instance, high levels of Pb, Cd, Zn, and Hg are commonly observed in carpophores of macrofungi in polluted areas. Accumulation of Au and Sb has also been observed in some fungal species, demonstrating a species-dependent accumulation. In addition to this, in controlled experiments, bioaccumulation in fruiting bodies of Agaricus bisporus has also been shown for Ag. As a result, the ability to accumulate metals in fruiting bodies of macrofungi has been proposed as an easy way to alleviate metal pollution in soils. However, to the best of our knowledge, this aspect has not yet been investigated to recover metals from e-waste.

References

- Toxin, science: britannica.com, Retrieved 12 May, 2020
- Toxicants-and-their-classification-toxicology, toxicology-toxicants-4486: environmentalpollution.in, Retrieved 27 August, 2020
- Environmental-toxicology, science: britannica.com, Retrieved 10 January, 2020
- All-about-environmental-toxin: precisionnutrition.com, Retrieved 09 March, 2020
- Health-effects-toxic-chemicals, resources-whmis-what-you-need-to-know: ehs.utoronto.ca, Retrieved 18 July, 2020
- Bioaccumulation, pharmacology-toxicology-and-pharmaceutical-science: sciencedirect.com, Retrieved 05 February, 2020

2

Sources of Environmental Toxicity

There are many sources of environmental toxicity that can lead to the presence of toxicants in our food, water and air. Some of them are polychlorinated biphenyl, lead, mercury, chromium, benzene, formaldehyde, radon, dichlorodiphenyltrichloroethane, etc. This chapter has been carefully written to provide an easy understanding of these sources of environmental toxicity.

Polychlorinated Biphenyls

Polychlorinated biphenyls (PCBs) are a group of manmade chemicals. They are oily liquids or solids, clear to yellow in color, with no smell or taste. PCBs are very stable mixtures that are resistant to extreme temperature and pressure. PCBs were used widely in electrical equipment like capacitors and transformers. They also were used in hydraulic fluids, heat transfer fluids, lubricants, and plasticizers. The primary company that made PCBs in the United States was Monsanto Inc., mainly using the name Aroclor. Commercial production of PCBs ended in 1977 because of health effects associated with exposure. In 1979, the U.S. Environmental Protection Agency (USEPA) banned the use of PCBs; however, PCBs are still present in many pre-1979 products.

PCBs have been released into the environment through spills, leaks from electrical and other equipment, and improper disposal and storage. It is estimated that more than half of the PCBs produced have been released into the environment. Once in the environment, PCBs can be transported long distances and they bind strongly to soil and sediment so they tend to be persistent in the environment. They have been found in air, water, soil, and sediments throughout the world. Because PCBs have not been made since 1977, the levels in the environment and in the food chain have been declining.

PCB Exposure

Since PCBs are found throughout the environment, it is likely that everyone has been exposed to them. PCBs can enter the body by eating or drinking contaminated food, through the air we breathe, or by skin contact. PCBs are easily absorbed by the body and are stored in fatty tissue. PCBs are not eliminated well, so they can accumulate in the body.

Most people are exposed to PCBs by eating contaminated fish, meat, and dairy products. Catfish, buffalo fish, and carp usually have the highest PCB levels. Plants take up only small amounts of PCBs from the soil, so amounts in grazing animals and dairy products are generally lower than in fish. Dust contaminated with very small levels of PCBs may be found on the outer surfaces of fruits and vegetables.

PCBs do not dissolve well, so exposure to them from water is usually not a concern. Some private wells may use old submersible pumps that contain PCB oil. If the pump seal fails, PCBs can leak into the well and contaminate the drinking water.

Older fluorescent lights found in schools, offices and homes may still contain transformers or ballasts that contain PCBs. If the ballasts fail, PCBs can leak out and contaminate exposed surfaces and the air. PCB levels measured in the air after a light ballast failure can be significant. The workplace also may be a source of PCB exposure.

Health Effects

Coming in contact with PCBs does not mean you will get sick or have health problems. Getting sick from being exposed to PCBs depends on: the amount of PCBs that entered your body, how long you were exposed to PCBs, and how sensitive your body is to PCBs.

In people, PCBs can affect the skin and may cause *chloracne*–small, pale, yellow skin lesions that may last from weeks to years. PCBs also can cause short-term changes in the activity of the liver, but without any noticeable symptoms. These liver changes are similar to those resulting from the consumption of alcoholic beverages or smoking cigarettes. Animal studies also have suggested that PCBs can affect the immune, endocrine and reproductive systems, but these effects are uncertain in humans.

Large amounts of PCBs given to laboratory animals over a short time can cause cancer. Studies of human workers exposed to high levels of PCBs for long periods have not consistently shown that PCBs cause cancer in humans. USEPA classifies PCBs as *probable*human carcinogens (cancer-causing chemicals), but there is no evidence that PCBs cause cancer at the low levels found in the environment.

Birth defects have been linked to mothers who have been exposed to PCBs. Developing fetuses and young children are the most vulnerable to PCBs, therefore, children and women who may become pregnant, are pregnant, or nursing should limit their exposure to PCBs. A pregnant woman can pass these chemicals to her unborn child. Mothers who eat highly contaminated fish before giving birth may have children who have slower mental development. PCBs also can be passed to a baby through breast milk. However, the significant benefits of breastfeeding far outweigh the risks. Young children also may experience developmental health effects.

Most people have a measurable amount of PCBs in their bodies. A blood test is best for measuring exposure to large amounts of PCBs. Although measuring PCBs in the body is possible, the analysis is expensive and not generally recommended because the results do not predict health effects or treatment.

Metal Toxicity

Metal toxicity or metal poisoning is the toxic effect of certain metals in certain forms and doses on life. Some metals are toxic when they form poisonous soluble compounds. Certain metals have no biological role, i.e. are not essential minerals, or are toxic when in a certain form. In the case of lead, any measurable amount may have negative health effects. Often heavy metals are thought as synonymous, but lighter metals may also be toxic in certain circumstances, such as beryllium and lithium. Not all heavy metals are particularly toxic, and some are essential, such as iron. The definition may also include trace elements when in abnormally high doses may be toxic. An option for treatment of metal poisoning may be chelation therapy, which is a technique which involves the administration of chelation agents to remove metals from the body.

Toxic metals sometimes imitate the action of an essential element in the body, interfering with the metabolic process resulting in illness. Many metals, particularly heavy metals are toxic, but some heavy metals are essential, and some, such as bismuth, have a low toxicity. Most often the definition of toxic metals includes at least cadmium, manganese, lead, mercury and the radioactive metals. Metalloids (arsenic, polonium) may be included in the definition. Radioactive metals have both radiological toxicity and chemical toxicity. Metals in an oxidation state abnormal to the body may also become toxic: chromium(III) is an essential trace element, but chromium(VI) is a carcinogen.

Toxicity is a function of solubility. Insoluble compounds as well as the metallic forms often exhibit negligible toxicity. The toxicity of any metal depends on its ligands. In some cases, organometallic forms, such as methylmercury and tetraethyl lead, can be extremely toxic. In other cases, organometallic derivatives are less toxic such as the cobaltocenium cation.

Decontamination for toxic metals is different from organic toxins: because toxic metals are elements, they cannot be destroyed. Toxic metals may be made insoluble or collected, possibly by the aid of chelating agents, or through bioremediation. Alternatively, they can be diluted into a sufficiently large reservoir, such as the sea, because immediate toxicity is a function of concentration rather than amount.

Toxic metals can bioaccumulate in the body and in the food chain. Therefore, a common characteristic of toxic metals is the chronic nature of their toxicity. This is particularly notable with radioactive heavy metals such as radium, which imitates calcium to the point of being incorporated into human bone, although similar health implications are found in lead or mercury poisoning.

Testing for Poisoning

People are continually exposed to metals in the environment. Medical tests can detect metals often, but this is to be expected and alone is not evidence that a person is poisoned. Metal screening tests should not be used unless there is reason to believe that a person has had excessive exposure to metals. People should seek medical testing for poisoning only if they are concerned for a particular reason, and physicians should consider a patient's history and physical examination before conducting tests to detect metals.

Treatment for Poisoning

Chelation therapy is a medical procedure that involves the administration of chelating agents to remove heavy metals from the body. Chelating agents are molecules that have multiple electron-donating groups, which can form stable coordination complexes with metal ions. Complexation prevents the metal ions from reacting with molecules in the body, and enable them to be dissolved in blood and eliminated in urine. It should only be used in people who have a diagnosis of metal intoxication. That diagnosis should be validated with tests done in appropriate biological samples.

Chelation therapy is administered under very careful medical supervision due to various inherent risks. Even when the therapy is administered properly, the chelation drugs can have significant side effects. Chelation administered inappropriately can cause neurodevelopmental toxicity, increase risk of developing cancer, and cause death; chelation also removes essential metal elements and requires measures to prevent their loss.

Specific Types of Poisoning

Aluminium Phosphide Poisoning

Aluminium has no known biological role and its classification into toxic metals is controversial.

Acute aluminium phosphide poisoning (AAlPP) is a large, though under-reported, problem in the Indian subcontinent. Aluminium phosphide (AlP), which is readily available as a fumigant for stored cereal grains, sold under various brand names such as QuickPhos and Celphos, is highly toxic, especially when consumed from a freshly opened container. Death results from profound shock, myocarditis and multi-organ failure. Aluminium phosphide has a fatal dose of between 0.15 and 0.5 grams (0.0053 and 0.0176 oz). It has been reported to be the most common cause of suicidal death in North India. The very high toxicity of aluminium phosphide is attributed to the phosphine content and is not related to aluminium. Calcium phosphide and zinc phosphide are similar poisons.

Arsenic Poisoning

Arsenic poisoning is a medical condition caused by elevated levels of arsenic in the body. The dominant basis of arsenic poisoning is from ground water that naturally contains high concentrations of arsenic. A 2007 study found that over 137 million people in more than 70 countries are probably affected by arsenic poisoning from drinking water.

Beryllium Poisoning

Beryllium poisoning is illness resulting from the toxic effect of beryllium in its elemental form or in various chemical compounds. The toxicity of beryllium depends upon the duration, intensity and frequency of exposure (features of dose), as well as the form of beryllium and the route of exposure (i.e. inhalation, dermal, ingestion). According to the International Agency for Research on Cancer (IARC), beryllium and beryllium compounds are Category 1 carcinogens; they are carcinogenic to both animals and humans.

Cadmium Poisoning

Cadmium is an extremely toxic metal commonly found in industrial workplaces. Due to its low permissible exposure limit, overexposures may occur even in situations where trace quantities of cadmium are found. Cadmium is used extensively in electroplating, although the nature of the operation does not generally lead to overexposures. Cadmium is also found in some industrial paints and may represent a hazard when sprayed. Operations involving removal of cadmium paints by scraping or blasting may pose a significant hazard. Cadmium is also present in the manufacturing of some types of batteries. Exposures to cadmium are addressed in specific standards for the general industry, shipyard employment, construction industry, and the agricultural industry.

Copper Toxicity

Copper toxicity, also called copperiedus, refers to the consequences of an excess of copper in the body. Copperiedus can occur from eating acid foods cooked in uncoated copper cookware, or from exposure to excess copper in drinking water, as a side-effect of estrogen birth control pills, or other environmental sources. It can also result from the genetic condition Wilson's disease.

Iron Poisoning

Iron poisoning is an iron overload caused by a large excess of iron intake and usually refers to an acute overload rather than a gradual one. The term has been primarily associated with young children who consumed large quantities of iron supplement pills, which resemble sweets and are widely used, including by pregnant women (approximately 3 grams is lethal for a 2 year old). Targeted packaging restrictions in the US for supplement containers with over 250 mg elemental iron have existed since 1978, and recommendations for unit packaging have reduced the several iron poisoning fatalities per year to almost nil since 1998. No known cases of iron poisoning have been identified that are associated with iron mining.

Lead Poisoning

Lead poisoning is a medical condition in humans and other vertebrates caused by increased levels of the heavy metal lead in the body. Lead interferes with a variety of body processes and is toxic to many organs and tissues including the heart, bones, intestines, kidneys, and reproductive and nervous systems. It interferes with the development of the nervous system and is therefore particularly toxic to children, causing potentially permanent learning and behavior disorders. Symptoms include abdominal cramping, constipation, tremors, mood changes, infertility, anemia, and toxic psychosis.

Lithium Poisoning

Lithium is used in some medications, specifically to treat bipolar disorder. The level of "sufficient" medication is thought by many physicians to be close to toxic tolerance for kidney function. Therefore, the patient is often monitored for this purpose.

Manganese Poisoning or Manganism

Manganism or manganese poisoning is a toxic condition resulting from chronic exposure to manganese and first identified in 1837 by James Couper.

Mercury Poisoning

Mercury poisoning is a disease caused by exposure to mercury or its compounds. Mercury (chemical symbol Hg) is a heavy metal occurring in several forms, all of which can produce toxic effects in high enough doses. Its zero oxidation state Hg^0 exists as vapor or as liquid metal, its mercurous state Hg_2^{2+} exists as inorganic salts, and its mercuric state Hg^{2+} may form either inorganic salts or organomercury compounds; the three groups vary in effects. Toxic effects include damage to the brain, kidney, and lungs. Mercury poisoning can result in several diseases, including acrodynia (pink disease), Hunter-Russell syndrome, and Minamata disease.

Symptoms typically include sensory impairment (vision, hearing, speech), disturbed sensation and a lack of coordination. The type and degree of symptoms exhibited depend upon the individual toxin, the dose, and the method and duration of exposure.

Silver Poisoning

Argyria or argyrosis is a condition caused by inappropriate exposure to chemical compounds of the element silver, or to silver dust. The most dramatic symptom of argyria is that the skin turns blue or bluish-grey. It may take the form of generalized argyria or local argyria. Generalized argyria affects large areas over much of the visible surface of the body. Local argyria shows in limited regions of the body, such as patches of skin, parts of the mucous membrane or the conjunctiva.

Thallium Poisoning

Thallium and its compounds are often highly toxic. Contact with skin is dangerous, and adequate ventilation should be provided when melting this metal. Many thallium(I) compounds are highly soluble in water and are readily absorbed through the skin. Exposure to them should not exceed 0.1 mg per m^2 of skin in an 8-hour time-weighted average (40-hour work week). Thallium is a suspected human carcinogen.

Tin Poisoning

Tin poisoning refers to the toxic effects of tin and its compounds. Cases of poisoning from tin metal, its oxides, and its salts are "almost unknown"; on the other hand certain organotin compounds are almost as toxic as cyanide.

Zinc Toxicity

Even though zinc is an essential requirement for a healthy body, excess zinc can be harmful, and cause zinc toxicity. Such toxicity levels have been seen to occur at ingestion of greater than 225 mg of Zinc. Excessive absorption of zinc can suppress copper and iron absorption. The free zinc ion in solution is highly toxic to bacteria, plants, invertebrates, and even vertebrate fish.

Arsenic

Arsenic is a natural component of the earth's crust and is widely distributed throughout the environment in the air, water and land. It is highly toxic in its inorganic form.

People are exposed to elevated levels of inorganic arsenic through drinking contaminated water, using contaminated water in food preparation and irrigation of food crops, industrial processes, eating contaminated food and smoking tobacco.

Long-term exposure to inorganic arsenic, mainly through drinking-water and food, can lead to chronic arsenic poisoning. Skin lesions and skin cancer are the most characteristic effects.

Drinking-water and Food

The greatest threat to public health from arsenic originates from contaminated groundwater. Inorganic arsenic is naturally present at high levels in the groundwater of a number of countries, including Argentina, Bangladesh, Chile, China, India, Mexico, and the United States of America. Drinking-water, crops irrigated with contaminated water and food prepared with contaminated water are the sources of exposure.

Fish, shellfish, meat, poultry, dairy products and cereals can also be dietary sources of arsenic, although exposure from these foods is generally much lower compared to exposure through contaminated groundwater. In seafood, arsenic is mainly found in its less toxic organic form.

Industrial Processes

Arsenic is used industrially as an alloying agent, as well as in the processing of glass, pigments, textiles, paper, metal adhesives, wood preservatives and ammunition. Arsenic is also used in the hide tanning process and, to a limited extent, in pesticides, feed additives and pharmaceuticals.

Tobacco

People who smoke tobacco can also be exposed to the natural inorganic arsenic content of tobacco because tobacco plants can take up arsenic naturally present in the soil. Also, in the past, the potential for elevated arsenic exposure was much greater when tobacco plants used to be treated with lead arsenate insecticide.

Health Effects

Inorganic arsenic is a confirmed carcinogen and is the most significant chemical contaminant in drinking-water globally. Arsenic can also occur in an organic form. Inorganic arsenic compounds (such as those found in water) are highly toxic while organic arsenic compounds (such as those found in seafood) are less harmful to health.

Acute Effects

The immediate symptoms of acute arsenic poisoning include vomiting, abdominal pain and diarrhoea. These are followed by numbness and tingling of the extremities, muscle cramping and death, in extreme cases.

Long-term Effects

The first symptoms of long-term exposure to high levels of inorganic arsenic (for example, through drinking-water and food) are usually observed in the skin, and include pigmentation changes, skin lesions and hard patches on the palms and soles of the feet (hyperkeratosis). These occur after a minimum exposure of approximately five years and may be a precursor to skin cancer.

In addition to skin cancer, long-term exposure to arsenic may also cause cancers of the bladder and lungs. The International Agency for Research on Cancer (IARC) has classified arsenic and arsenic compounds as carcinogenic to humans, and has also stated that arsenic in drinking-water is carcinogenic to humans.

Other adverse health effects that may be associated with long-term ingestion of inorganic arsenic include developmental effects, diabetes, pulmonary disease, and cardiovascular disease. Arsenic-induced myocardial infarction, in particular, can be a significant cause of excess mortality. In China (Province of Taiwan), arsenic exposure has been linked to "Blackfoot disease", which is a severe disease of blood vessels leading to gangrene. This disease has not been observed in other parts of the world however, and it is possible that malnutrition contributes to its development.

Arsenic is also associated with adverse pregnancy outcomes and infant mortality, with impacts on child health , and exposure in utero and in early childhood has been linked to increases in mortality in young adults due to multiple cancers, lung disease, heart attacks, and kidney failure. Numerous studies have demonstrated negative impacts of arsenic exposure on cognitive development, intelligence, and memory.

Magnitude of the Problem

Arsenic contamination of groundwater is widespread and there are a number of regions where arsenic contamination of drinking-water is significant. It is now recognized that at least 140 million people in 50 countries have been drinking water containing arsenic at levels above the WHO provisional guideline value of 10 μg/L.

Arsenic in Bangladesh has attracted much attention since recognition in the 1990s of its wide occurrence in well-water in that country. Since this time, significant progress has since been made and the number of people exposed to arsenic exceeding the Bangladesh drinking-water quality standard has decreased by approximately 40%. Despite these efforts, it was estimated that in 2012 about 19 million and 39 million people in Bangladesh were still exposed to arsenic concentrations above the national standard of 50 μg/L and the WHO provisional guideline value of 10 μg/L respectively. In a highly affected area of Bangladesh, 21.4% of all deaths in the area were attributed to arsenic levels above 10 μg/L in drinking-water. A similar dose-response function has been found in other parts of Bangladesh, and these these results have been combined with national survey data to estimate an annual death toll of nearly 43 000. The US National Research Council has noted that as many as 1 in 100 additional cancer deaths could be expected from a lifetime exposure to drinking-water containing 50 μg/L.

The symptoms and signs caused by long-term elevated exposure to inorganic arsenic differ between individuals, population groups and geographical areas. Thus, there is no universal definition of the disease caused by arsenic. This complicates the assessment of the burden on health of arsenic.

Similarly, there is no method to distinguish cases of cancer caused by arsenic from cancers induced by other factors. As a result, there is no reliable estimate of the magnitude of the problem worldwide.

In 2010, the Joint FAO/WHO Expert Committee on Food Additives (JECFA) re-evaluated the effects of arsenic on human health, taking new data into account. JECFA concluded that for certain regions of the world where concentrations of inorganic arsenic in drinking-water exceed 50–100 μg/L, there is some evidence of adverse effects. In other areas, where arsenic concentrations in water are elevated (10–50 μg/L), JECFA concluded that while there is a possibility of adverse effects, these would be at a low incidence that would be difficult to detect in epidemiological studies.

Prevention and Control

The most important action in affected communities is the prevention of further exposure to arsenic by the provision of a safe water supply for drinking, food preparation and irrigation of food crops. There are a number of options to reduce levels of arsenic in drinking-water.

Substitute high-arsenic sources, such as groundwater, with low-arsenic, microbiologically safe sources such as rain water and treated surface water. Low-arsenic water can be used for drinking, cooking and irrigation purposes, whereas high-arsenic water can be used for other purposes such as bathing and washing clothes.

Discriminate between high-arsenic and low-arsenic sources. For example, test water for arsenic levels and paint tube wells or hand pumps different colours. This can be an effective and low-cost means to rapidly reduce exposure to arsenic when accompanied by effective education.

Blend low-arsenic water with higher-arsenic water to achieve an acceptable arsenic concentration level.

Install arsenic removal systems – either centralized or domestic – and ensure the appropriate disposal of the removed arsenic. Technologies for arsenic removal include oxidation, coagulation-precipitation, absorption, ion exchange, and membrane techniques. There is an increasing number of effective and low-cost options for removing arsenic from small or household supplies, though there is still limited evidence about the extent to which such systems are used effectively over sustained periods of time.

Long-term actions are also required to reduce occupational exposure from industrial processes. Education and community engagement are key factors for ensuring successful interventions. There is a need for community members to understand the risks of high arsenic exposure and the sources of arsenic exposure, including the intake of arsenic by crops (e.g. rice) from irrigation water and the intake of arsenic into food from cooking water. High-risk populations should also be monitored for early signs of arsenic poisoning – usually skin problems.

Lead

Lead toxicity is an important environmental disease and its effects on the human body are devastating. There is almost no function in the human body which is not affected by lead toxicity. Though in countries like US and Canada the use of lead has been controlled up to a certain extent, it is still used vehemently in the developing countries. This is primarily because lead bears unique physical and chemical properties that make it suitable for a large number of applications for which humans have exploited its benefits from historical times and thus it has become a common environmental pollutant. Lead is highly persistent in the environment and because of its continuous use its levels rise in almost every country, posing serious threats.

Lead is the most important toxic heavy element in the environment. Due to its important physico-chemical properties, its use can be retraced to historical times. Globally it is an abundantly distributed, important yet dangerous environmental chemical. Its important properties like softness, malleability, ductility, poor conductibility and resistance to corrosion seem to make difficult to give up its use. Due to its non-biodegradable nature and continuous use, its concentration accumulates in the environment with increasing hazards.

Human exposure to lead and its compounds occurs mostly in lead related occupations with various sources like leaded gasoline, industrial processes such as smelting of lead and its combustion, pottery, boat building, lead based painting, lead containing pipes, battery recycling, grids, arm industry, pigments, printing of books, *etc.*

Though its widespread use has discontinued in many countries of the world, it is still used in many industries like car repair, battery manufacturing and recycling, refining, smelting, *etc.* Lead is a highly poisonous metal affecting almost every organ in the body. Of all the organs, the nervous system is the mostly affected target in lead toxicity, both in children and adults. The toxicity in children is however of a greater impact than in adults. This is because their tissues, internal as well as external, are softer than in adults. Long-term exposure of adults can result in decreased performance in some tests of cognitive performance that measure functions of the nervous system. Infants and young children are especially sensitive to even low levels of lead, which may contribute to Behavioral problems, learning deficits and lowered IQ. Long-time exposure to lead has been reported to cause anaemia, along with an increase in blood pressure, and that mainly in old and middle aged people. Severe damage to the brain and kidneys, both in adults and children, were found to be linked to exposure to heavy lead levels resulting in death. In pregnant women, high exposure to lead may cause miscarriage. Chronic lead exposure was found to reduce fertility in males. Blood disorders and damage to the nervous system have a high occurrence in lead toxicity.

Detection of Lead Poisoning

Several methods are used to detect elevated blood lead levels. The presence of changes in blood cells visible under the microscope or deletion of dense lines in the bones of children seen on X-ray are signs used for detecting lead poisoning. However the main tool to detect elevated levels of body lead is to measure the level of lead in blood samples. This test gives however only an account of lead present in circulating blood but cannot show how much lead is stored in the body. As of 2012, the Centers for Disease Control and Prevention (USA) have set the standard elevated blood lead level for adults to be 10 μg/dL and for children 5 μg/dL of the whole blood. Previously, the standard lead level for children was 10 μg/dL. The appearance of clinical manifestations varies from individual to individual depending on other environmental factors. In some there is a clear appearance of clinical features even at lower levels, while some are asymptomatic even at higher levels of lead present in their body fluids. Children are more prone to the effects of lead because usually their organs are in a developing stage. Thus blood lead levels have to be set lower and must be frequently checked, particularly where contamination is expected.

Effects of Lead Poisoning

All along human history, lead poisoning has been reported to have severe effects. Occasional lead poisoning was found to be caused by lead salts used in pottery glazes leached by acidic fruit juices. Beethoven's death has been treated in various reports. Many of them have concluded that he died because of the toxic doses of lead-based treatment administered by his doctor. Analysis of his hair was found to contain elevated levels of lead. It is also assumed that in the eighteenth and early nineteenth century lead was illegally added to wine both as a sweetener and to make it appear fresh. Lead poisoning is believed to be primarily responsible for the collapse of the Roman Empire, in which lead acetate was used as a sweetener of wine. Its prolonged use was considered to have caused dementia to many Roman emperors. Lead poisoning has also been found to be the cause of anaemia in a number of cases as lead inhibits porphobilinogen synthase and ferrochelatase, preventing both porphobilinogen formation and the incorporation of iron into protoporphyrin IX, which prevents heme synthesis or causes ineffective heme synthesis and subsequently microcytic anaemia. One of the mechanisms by which lead interferes with cognition is that it acts as calcium analogue which interferes with ion channels. It has been observed that Pb^{2+} is a potent reversible and selective blocker of voltage-dependent calcium channels at low concentrations. The toxic effects on blood cells of rats caused by lead nitrate was alleviated by sodium selenite. They also showed that effects of lead nitrate were more harmful in diabetic than in non-diabetic rats. Oxidative stress was studied by low level lead exposure in first grade Uruguayan children, suggesting its potentially adverse effects on oxidative stress. Impaired respiratory function was observed in workers exposed to lead with elevated blood lead concentration and zinc protoporphyrin concentration.

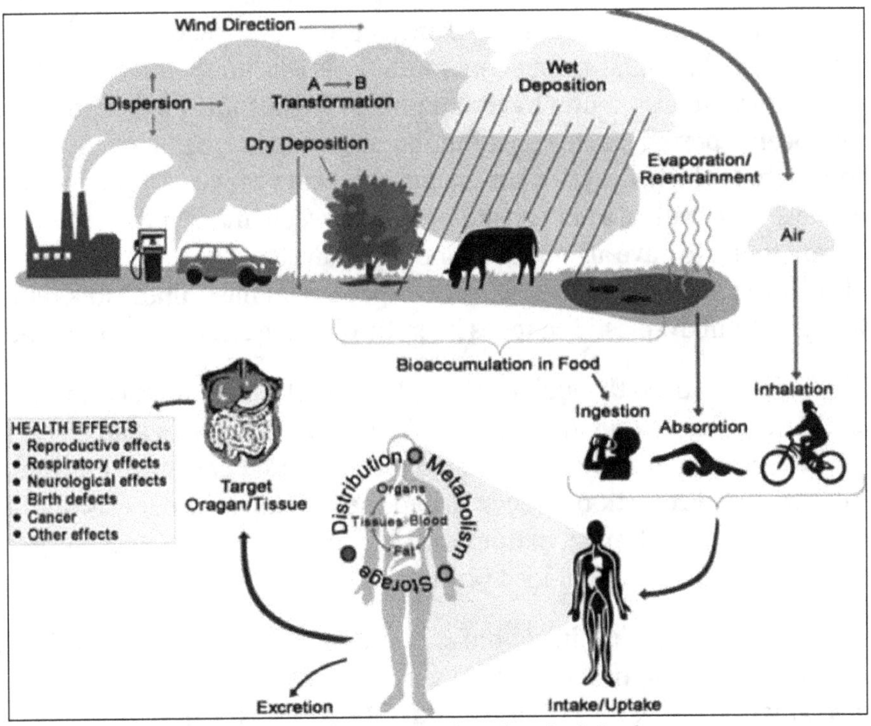

The people are exposed to chemicals in the environment and the effect of such chemicals on human health.

In a recent study lead and cadmium contents were investigated in counterfeit cigarettes seized in the United States by various law enforcement agencies. Both lead and cadmium levels were found to be markedly higher than in their genuine equivalents. The results suggest the possibility of a higher risk of serious consequences to public health caused by counterfeit cigarettes.

Possible Pathway of Lead Poisoning

Poisoning due to lead occurs mainly by ingestion of food or water contaminated with lead. However accidental ingestion of contaminated soil, dust or lead based paint may also result in poisoning. Lead is thought to be quickly absorbed in the blood stream and is believed to have adverse effects on certain organ systems like the central nervous system, the cardiovascular system, kidneys, and the immune system. Most pharmaceutical companies have set a limit for maximum daily intake of lead as 1.0 μg/g, however prolonged intake of even this low level of lead is hazardous to human beings. Occupational exposure also results in elevated blood lead levels. Increased blood levels are associated with delayed puberty in girls. There is no threshold value for the level of lead present in blood below which its concentration can be considered safe. Extremely low yet permanent levels of lead exposure were found to reduce the cognitive capacity of children. Dangers from lead poisoning due to paint pigments mainly in children have sharply reduced the use of lead in paints. However old houses may still contain substantial amounts of lead paint. This sometimes causes an accidental contamination of lead in children. In industrialised countries white lead paint has been completely withdrawn from sale, however yellow lead chromate is still in use. Old paint should not be stripped by sanding, as this produces inhalable dust. Traditional medicines were also found to contain heavy metals including lead. A number of diseases have been reported due to consumption of traditional medicine. Ayurvedic medicines are considered to be heavily contaminated with heavy metals. In one recent study the blood lead levels were evaluated in consumers of ayurvedic medicines. Of the 115 participants 40% were found to have an elevated blood lead levels of 10 μg/dL or above and 9.6% had blood lead levels above 50 μg/dL. Recently a patient taking Chinese folk remedies was reported to suffer from dysplastic changes in erythroid precursors due to lead poisoning.

Lead toxicity may be caused through fruits and vegetables contaminated with high lead levels from the soils where they were grown. The soil accumulates lead levels generally from pipes, lead paint and residual emissions from leaded gasoline that was used before the Environment Protection Agency issued the regulation around 1980. In order to prevent the general population from domestic lead poisoning, it is necessary to educate people about the major sources of lead poisoning.

Lead from water pipes coming into homes is one of the major sources. For ingested lead, the rate of absorption by the body is very high with almost 20–70% and in children it is even higher. However the rate of skin absorption for inorganic lead is low.

Signs and Symptoms

Lead poisoning causes a variety of symptoms, including abnormal Behavior which varies from person to person, while time of exposure plays an important role. There are also studies which show no symptoms of lead poisoning even with elevated levels of lead in the body. The question what makes such differences in the human body is an issue of major concern. Why are there such differences in attaining toxicity of lead or other toxic substances. Why for one group of people any increase in the concentration gives toxicity and for another the same level has no effect. Currently there is no evidence from the literature about such differences, however it is hoped that future research could explore this area and figure out the possible reasons. It usually takes weeks to months to build up the manifestations in the body by chronic lead exposure; however acute symptoms and signs also occur from short time intense exposures.

Organic lead is perhaps more toxic than inorganic lead because of its lipid soluble nature, which results in rapid consequences. However, the lead levels at which signs and symptoms appear vary widely, depending on unknown characteristics of each individual.

Blood lead levels from 25 and 60 µg/dL give rise to neuropsychiatric effects such as delayed reaction times, irritability, and difficulty in concentrating, as well as slowed down motor nerve conduction and headache. Anaemia may appear at blood lead levels higher than 50 µg/dL. In adults, abdominal colic, involving paroxysms of pain, may appear at blood lead levels higher than 80 µg/dL. High blood lead levels which exceed 100 µg/dL cause very severe manifestations, like signs of encephalopathy (condition characterized by brain swelling) accompanied by increased pressure within the skull, delirium, coma, seizures, and headache. However such manifestations appear in children at lead levels of 70 µg/dL and more.

Central nervous system and neuromuscular manifestations usually result from intense exposure, while gastrointestinal features usually result from exposure over longer periods. Signs and symptoms of chronic exposure include loss of short-term memory or concentration, depression, nausea, abdominal pain, loss of coordination, and numbness and tingling in the extremities. Fatigue, problems with sleep, headaches, stupor, slurred speech, and anaemia are also found in chronic lead poisoning. Children with chronic poisoning generally show aggressive Behavior and refuse to play.

Effects on Children

Pregnant women who have elevated blood lead levels are at a risk of premature birth or of babies with a low birth weight. The foetus may be adversely affected at blood lead concentrations well below 25 µg per deciliter. Blood lead levels in the neonate were found to be higher than simultaneous maternal lead levels. Emaciated women with substantial exposure to lead prior to pregnancy are considered to be at increased risk.

Children have been repeatedly reported to be at higher risk for lead poisoning because their bodies are in a state of growth and development. Moreover, the absorption of lead occurs more quickly in children than in adults. Children, due to their childish Behavior, are more prone to ingest and inhale dust contaminated with lead.

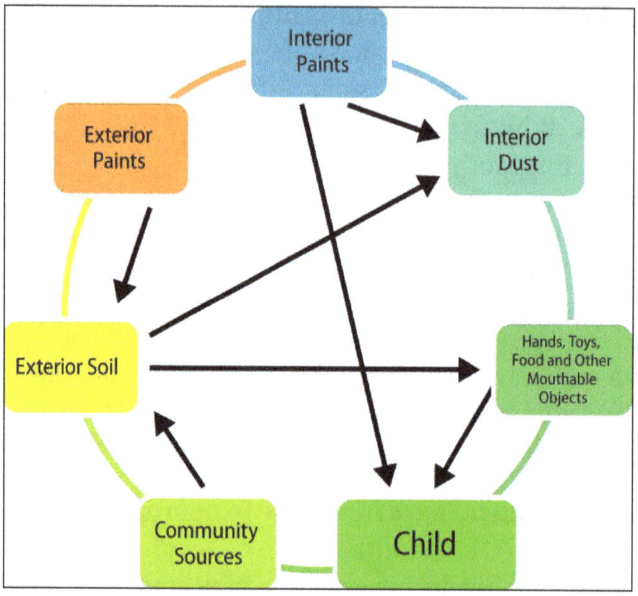

How children become easy targets for lead poisoning in the environment home.

Pathophysiology

Exposure occurs through various ways like inhalation, ingestion or skin contact. Direct contact of lead or lead-based compounds through the mouth, nose, eyes and through cracks of the skin may also increase lead levels. In adults, about 35–40% of inhaled lead dust is deposited in the lungs and about 95% goes into the bloodstream. On ingestion of inorganic lead almost 15% is absorbed, however this value is higher in children, pregnant women and people with deficiencies of calcium, zinc or iron. A certain amount of the lead which is generally bound in tissues like bones, teeth, hair or nails is considered to be nontoxic because of its unavailability to other tissues. The rate of absorption of lead in bones and teeth is high amounting to almost 94% in adults, while in children this rate is 70%, which allows the soft tissues to absorb more lead and thus to cause serious health consequences. The half-life of lead in such tissues results in its induction into the blood stream long after the initial exposure. Blood lead has a lower half-life of just about 40 days in humans. This increases in the case of pregnant women and of children whose bones are in a developing stage. The developing bones in children which undergo remodelling allow the lead to be continuously reintroduced into the blood stream.

Due to a prolonged exposure of lead for years, a much slower clearance takes place. This is due to prolonged accumulation of lead in bones released over a long period of time. Along with bones, teeth and blood, many other tissues store lead in the body, *i.e.* the brain, spleen, kidneys, liver and lungs. Small amounts of lead were found to be removed through faeces and small amounts through hair, nails and sweat. And yet, the interesting point to include is that lead has been found to have no physiological role in the body, while its harmful effects are manifold. The effects of lead have been well studied also at cellular level. Heavy metals, including lead, create reactive radicals which damage cell structures, including DNA and cell membrane. Lead also interferes with the enzymes that help in the synthesis of vitamin D and with enzymes that maintain the integrity of the cell membrane. Lead was also found to interfere with DNA transcription.

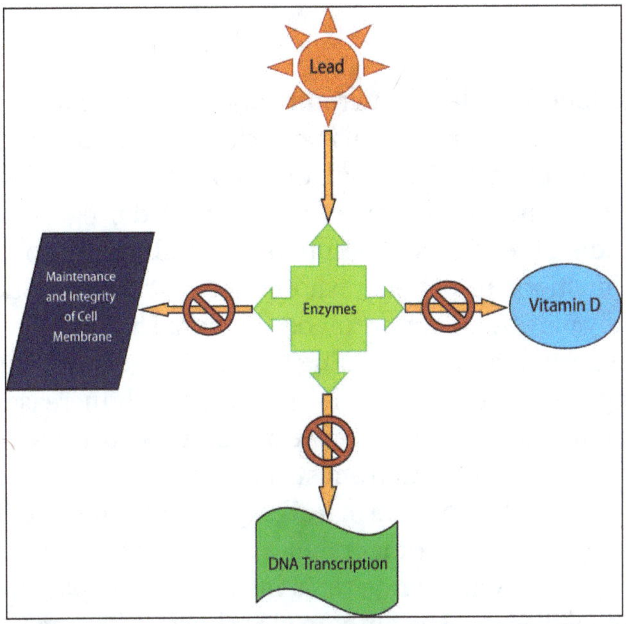

Illustration of the action of lead on enzymes, leading to the disruption of vitamin D synthesis, maintenance of cell membrane and DNA transcription.

As lead disrupts the maintenance of the cell membrane, red blood cells with a damaged membrane become more fragile, which results in anaemia. Lead is also speculated to alter the permeability of blood vessels and collagen synthesis. Damaged activity of cells of the immune system, such as polymorphonuclear leukocytes, results in decreasing immune activity.

One of the main reasons for lead poisoning causing anaemia is that lead interferes with the activity of an essential enzyme called delta-aminolevulinic acid dehydratase, or ALAD, which is important in the biosynthesis of heme, the cofactor found in haemoglobin. Heme precursors, such as aminolevulinic acid, have been found to build up due to their interference with lead, which may be directly or indirectly harmful to neurons. In a recent finding accumulation of organic lead in the liver was found to induce oxidative imbalance and protein impairment that may result in ER stress followed by liver injuries. In another recent study the effect of lead exposure on C57BL/6J mice was observed. Early chronic low-level lead exposure yielding BLLs ranging from 1.98 to 14.84 μg/dL was found to disrupt exploratory activity in pre-adolescent mice. In another recent study the authors studied the comorbid effects of lead exposure and high fat diet (HFD). They observed that both lead and HFD were involved in suppressing osteoblastogenesis and altered progenitor cell differentiation, promoting osteoclastogenesis and increasing adipogenesis. In addition, an increased *in vitro* PPAR-γ activity and inhibited β-catenin activity were observed in MC3T3 cells due to lead and non esterified fatty acids. They further concluded that lead and HFD were involved in producing selective deficit bone accrual in which reduced Wnt signalling may be involved with associations in progenitor cell activity alteration.

Renal System

A number of studies found low level environmental lead exposure to be associated with accelerated deterioration of chronic renal insufficiency. Even at levels far below the limits of normal ranges in the general population, both blood lead level and blood lead burden were found to be increased, predicting accelerated progression of chronic renal disease. Wedeen studied the effects of occupational lead hazards on patients. They observed tubular dysfunctions in patients who underwent biopsy. Such studies suggest that lead nephropathy may be an important occupational hazard. Excretion of the waste product urate was reported in lead poisoning, suggesting gout, in which urate builds up in the body. Prolonged exposure to lead is associated with increase in blood pressure and some studies showed a connection between lead exposure and coronary heart disease, heart rate variability, and death from stroke, but this evidence is limited. Chronic lead nephropathy occurred due to years of lead exposure manifested in kidney biopsy by moderate focal atrophy, loss of proximal tubules and interstitial fibrosis. Low level environmental lead exposure may accelerate renal insufficiency in patients without diabetes who have chronic renal disease. The authors also showed that repeated chelation therapy may improve renal function and slow down the progression of renal insufficiency.

Reproductive System

The reproductive system of both males and females is affected by lead. In males sperm count is reduced and other changes occur in the volume of sperm when blood lead levels exceed 40 μg/dL. Activities like motility and the general morphology of sperm are also affected at this level. The problems with the reproductivity of females due to lead exposure are more severe. Toxic levels of lead can lead to miscarriages, prematurity, low birth weight, and problems with development during childhood. Blood lead levels in mothers and infants are usually similar as the lead present in mother blood passes into the foetus through the placenta and also through breast milk. As bones store the highest content of lead, if metabolic changes mobilise the lead from bones into the blood due to pregnancy, the lead toxic risks increase. However increased calcium intake during pregnancy can help mitigate this phenomenon. Low doses of lead were found to significantly reduce the number of sperms within the epididymis of mice, while high doses reduced both the sperm count and percentage of motile sperms and led to an increased percentage of epididymal abnormal sperms. Lead directly targets testicular spermatogenesis and also the sperms in the epididymis inducing reproductive toxicity. In lead treated groups suppressed rates of serum testosterone, intratesticular sperm counts, and sperm production rates were reported. On investigating the reversibility of the toxic effects of lead on the male reproductive axis, it was shown that serum testosterone and sperm parameters normalised at the end of the recovery period in prepubertal animals yet not in pubertal animals. Prepubertal rats were shown to be less sensitive to the toxic effects of lead than rats whose exposure to lead was started after initiation of puberty. In rats exposed to lead, an 80% reduction in plasma and testicular testosterone and a 32% reduction in the plasma luteinizing hormone (LH) were reported. In lead exposed rats a sharp decline in the testosterone: LH ratio was observed. The detailed mechanism of how lead induces male infertility was reviewed. Some authors have suggested that a delay in the development of the sex specific pituitary growth hormone secretion pattern would possibly lead to growth effects of lead rather than to a persistent defect in development. Long-term low-dose lead exposure was shown to alter the signalling system between the hypothalamus and pituitary gland of male rats. This signalling is disrupted by long-term exposure, altering thereby the gonadotropin-releasing hormone system in the male rat. In a recent study by Zhang, the authors investigated lead interactions with human chorionic gonadotropin (HCG). Investigation of HCG by UV-vis absorption spectroscopy, circular dichroism spectroscopy, and ELISA indicated that lead acetate changed the secondary structure of HCG by loosening and destruction of the HCG skeleton and by increasing the hydrophobicity around Tyr residues, which resulted in decreased bioactivities of HCG. This work presents direct interactions of lead with sex hormones and suggests a possible mechanism of lead induced reproductive toxicity at molecular level.

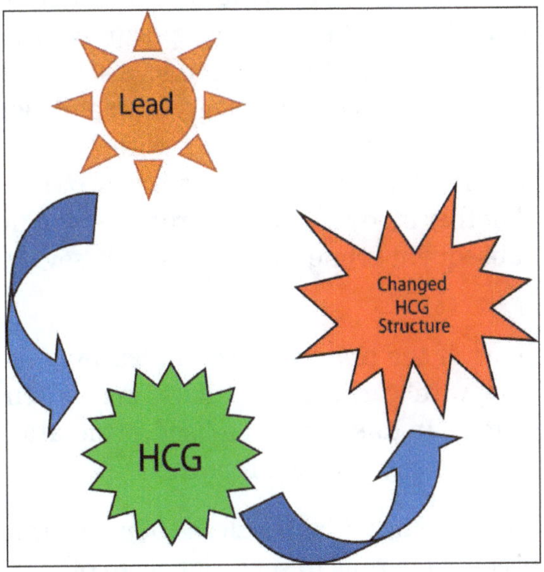

Depiction of lead changing the structure of HCG.

Nervous System

The brain is the most sensitive organ to lead exposure. In a child's developing brain, synapse formation is greatly affected in the cerebral cortex by lead. Lead also interferes with the development of neurochemicals, including neurotransmitters, and organisation of ion channels. Lead poisoning also causes loss of neuron myelin sheath, reduction in the number of neurons, it interferes with neurotransmission and decreases neuronal growth. The brain of adults exposed to increased lead levels during their childhood also shows a decreased volume, especially in the prefrontal cortex on MRI. Lead is able to pass through the endothelial cells at the blood brain barrier because it can substitute for calcium ions and be taken up by calcium-ATPase pumps, thereby interfering with synapse formation. Children with blood lead concentration greater than 10μg/dL are at higher risk for developmental disabilities.

The effect of lead on children´s cognitive abilities takes place at very low levels. There is apparently no lower threshold to the dose-response relationship below which lead exposure is treated as safe. Blood lead levels lower than 5 μg/dL were found to be associated with reduced academic performance Blood lead levels below 10 μg/dL were reported to be associated with lower IQ and Behavior problems such as aggression, in proportion with the given blood lead level. Between the blood lead levels of 5 and 35 μg/dL, an IQ decrease of 2–4 points for each μg/dL increase was reported in children. Increased blood lead levels are also associated with a decrease in cognitive performance and with other psychiatric conditions like depression and anxiety. An increase in the blood lead levels from 50 to about 100 μg/dL in adults was found to be associated with more severe conditions, like permanent impairment of central nervous system function. With increased lead exposure in children, an increase in neuropsychiatric disorders like attention deficit hyperactivity disorder and antisocial Behavior were found. Prenatal and early childhood lead exposure was reported in correlation with violent crimes in adulthood. The highest lead levels in the air were also shown to deviate normal behavior and turn to become aggressive and violent, thus for example the highest murder rates were found in countries with high levels of lead in the air. One study theorizes that lead exposure explains 65% to 90% of the variation in violent crime rates in the US. Another study showed a strong association between preschool blood lead levels and subsequent crime rate trends over several decades across nine countries.

The hippocampus is a part of the brain involved in learning and memory. The main reasons for lead interfering with learning particularly in children is that it damages the cells within the hippocampus. In rats exposed to lead, structural damage such as irregular nuclei and denaturation of myelin were reported.

Lead is also involved in interfering with the release of neurotransmitters. Neurotransmitters are chemicals used by neurons to send signals to other cells. This interference leads to the disruption of communication between cells. Lead usually interferes with the neurotransmitter glutamate which is important for many functions, like learning. It operates by blocking NMDA receptors. Blocking of such receptors is thought to be the main target of lead toxicity. One study found that in addition to inhibition of the NMDA receptor, lead exposure also decreased the amount of gene for this receptor in part of the brain. Lead was also found to be involved in apoptosis of brain cells in animal studies.

Mercury

Mercury exists naturally and as a man-made contaminant. The release of processed mercury can lead to a progressive increase in the amount of atmospheric mercury, which enters the atmospheric-soil-water distribution cycles where it can remain in circulation for years. Mercury poisoning is the result of exposure to mercury or mercury compounds resulting in various toxic effects depend on its chemical form and route of exposure. The major route of human exposure to methylmercury (MeHg) is largely through eating contaminated fish, seafood, and wildlife which have been exposed to mercury through ingestion of contaminated lower organisms. MeHg toxicity is associated with nervous system damage in adults and impaired neurological development in infants and children. Ingested mercury may undergo bioaccumulation leading to progressive increases in body burdens. This review addresses the systemic pathophysiology of individual organ systems associated with mercury poisoning. Mercury has profound cellular, cardiovascular, hematological, pulmonary, renal, immunological, neurological, endocrine, reproductive, and embryonic toxicological effects.

Mercury is one of the most toxic elements or substances on the planet to arsenic and lead that continues to be dumped into our waterways and soil, spilled into our atmosphere, and consumed in our food and water. Human activities have nearly tripled the amount of mercury in the atmosphere and the atmospheric burden is increasing 1.5 percent per year. Soil contaminated by mercury or the redistribution of contaminated water has the potential to enter the food chain through plant and livestock. Once in the food chain mercury can bioaccumulate causing adverse effects to human health. The exact mechanisms by which mercury enters the food chain remains largely unknown, and probably varies among ecosystems.

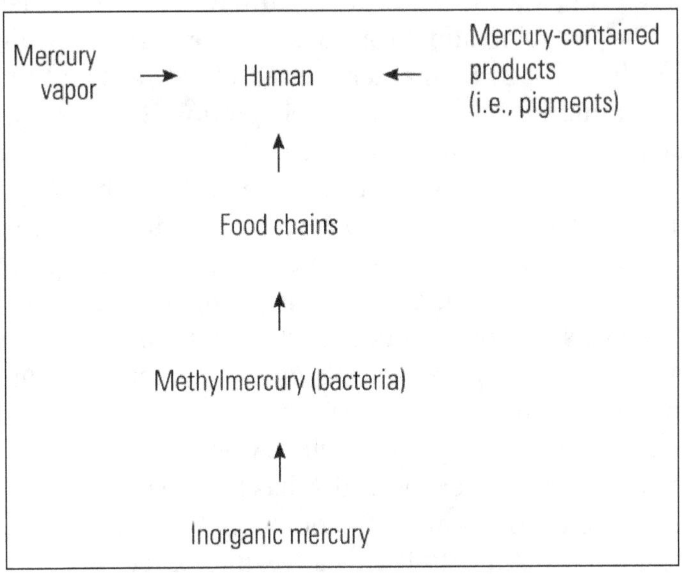

Routes of mercury exposure in humans.

Environmental mercury can exist in its elemental form, as inorganic mercury or as organic mercury. In its elemental form mercury exists as liquid metal, which in spite of its low vapor pressure (2 μm Hg), can be converted to a vapor at room temperature due to its low latent heat of evaporation (295 kJ/kg) and its relative absence from ambient air. Current sources of human exposure to elemental mercury included dental amalgam, thermometers, sphygmomanometer, barometers, fossil fuel emissions, incandescent lights, batteries, ritualistic practices using mercury, and the incineration of medical waste. Toxic vapors formed from mercury vaporization or the burning of mercury containing materials can enter the respiratory system and pass readily into the circulation. The average whole body biological half-life of inhaled mercury is approximately 60 days. Because mercury vapor can become lipid soluble once oxidized the potential exist for bioaccumulation in the renal cortex, liver, and especially the brain. It is estimated that the half-life of mercury in the brain can be as long as 20 years.

States of Mercury

Inorganic mercury exists in either the mercurous and mercuric form. Like oxidized elemental mercury, mercuric salts are more water soluble and toxic than elemental mercury. Mercuric salts are also easily absorbed by the gastrointestinal tract. The average whole body half-life of inorganic mercury is about 40 days.

The most common form of organic mercury is methylmercury (MeHg), which is the major source of organic mercury found in the ecosystems. MeHg is readily transported by water into the aquatic ecosystems. Because of its low water solubility it is considered to be relatively lipid soluble. MeHg is easily taken up by lower organisms, tends to work its way up the food chain and exhibits a proclivity to bioaccumulate in fish. Fish appear to be the primary source of MeHg poisoning in humans. Through mechanisms which are not yet known, various species of fish tend to have higher rates of MeHg bioaccumulation. The gastrointestinal tract absorbs approximately ninety five percent of ingested MeHg where it can then enter the red blood cells and the brain by binding covalently to glutathione and cysteine protein groups. Because urinary excretion of MeHg is negligible, MeHg is primarily eliminated from the body in an inorganic form through the action of the biliary system at the rate of 1% of the body burden per day. The biological half-life of MeHg is 39 to 70 days depending on body burden. Potential sources of organic mercury included exposure to fossil fuel emissions, the incineration of medical waste, dental amalgam, and various commercial products including skin creams, germicidal soaps, various medications, teething powders, analgesics, diaper treatments, vaccinations, thermometers, sphygmomanometer, barometers, incandescent lights, and batteries. Other sources for organic mercury include phenyl mercury compounds and ethyl mercury compounds, which were components of latex paints that were used before 1990s and thimerosal which has been used as a preservative in vaccines. Among the most dangerous mercury compound is dimethylmercury ($(CH_3)_2Hg$) which is toxic enough to cause death if only a few microliters is spilled on the skin, or even latex gloves. Mercury poisoning can result in death, mental retardation, dysarthria, blindness, neurological deficits, loss of hearing, developmental defects, and abnormal muscle tone.

Systemic Toxicological Effects of Mercury

Mercury exposure has been associated with the induction of over 250 symptoms which can complicate accurate diagnosis. Differential diagnosis begins with a patient history and physical examination consistent with mercury exposure. Laboratory testing typically includes: 1) Blood analysis; 2) Urinalysis, with a 24-hour urine analysis, and a urine challenge test with a "chelating" agent; 3) Hair analysis; and (d) Tissue biopsy if warranted. Because mercury can be quickly removed from the blood, redistributed and sequestered into different tissues it is important to note that there may not be a direct correlation between blood mercury concentration and the severity of mercury poisoning. Indeed, it is thought that shortly after entering the body that mercury quickly becomes tightly bound in the brain, spinal cord, ganglia, autonomic ganglia, and peripheral motor neurons. Nonetheless although the nervous system is the primary repository for mercury exposure, the transient and residual systemic distribution of mercury has the potential to cause symptoms in a number of different organ systems. In addition reports indicate that individual genetic background may play a role in mercury toxicokinetics.

Cellular Effects of Mercury

At the cellular level mercury exposure is associated with alterations in membrane permeability, changes in macromolecular structure due to its affinity for sulfhydryl and thiol groups, and DNA damage. Mercury has also been shown to induce oxidative stress and mitochondrial dysfunction which can result in alterations in calcium homeostasis and increased lipid peroxidation. In addition, mercury may also increase radical oxygen species levels because of its ability to act as a catalyst for Fenton-type reactions.

Cardiovascular, Hematological and Pulmonary Effects

Mercury accumulation in the heart is thought to contribute to cardiomyopathy. Indeed, mercury levels in the heart tissue of individuals who died from idiopathic dilated cardiomyopathy were found to be on average 22 000 times higher than in individuals who died of other forms of heart disease. Mercury poisoning may also cause chest pain or angina, especially in individuals under age 45. *In vitro* studies have indicated that MeHg can inhibit the cardioprotective activity of paraoxonase 1. There is also good evidence linking mercury with anemia including hemolytic anemia and aplastic anemia as mercury is thought to compete with iron for binding to hemoglobin which can result in impaired hemoglobin formation. In addition to anemia, additional data has also suggested that mercury may be a causative factor in mononucleosis and involved in leukemia, and Hodgkin's disease.

Toxic vapors formed from mercury vaporization or the burning of mercury containing materials can enter the respiratory system and pass readily into the circulation. Case control studies have demonstrated that the chronic inhalation of even low concentrations of mercury (0.7 to 42 $\mu g/m^3$) can produce tremors, sleep disturbances, and impaired cognitive skills in workers. Mercury poisoning is associated with several different pulmonary conditions including Young's syndrome, bronchitis and pulmonary fibrosis.

Effects on the Digestive and Renal Systems

Mercury is absorbed through the epithelial cells when ingested. This absorbed mercury can cause various digestive disturbances as it can inhibit the production of the digestive trypsin, chymotrypsin, and pepsin along with the function of xanthine oxidase and dipeptyl peptidase IV. The effects of mercury on the gastrointestinal system typically present as abdominal pain, indigestion, inflammatory bowel disease, ulcers and bloody diarrhea. Mercury ingestion has also been associated with the destruction of intestinal flora which can increase the amount of undigested food products in the blood stream causing immune mediated reactions and reduced resistance to pathogenic infection.

Mercury can cause kidney damage and evidence suggests a linkage between mercury exposure and acute tubular necrosis, glomerulonephritis, chronic renal disease, renal cancer and nephrotic syndrome. Various reports have shown mercury exposure can lead to various kidney injuries including: subacute-onset nephrotic syndrome, tubular dysfunction, secondary focal segmental glomerulosclerosis, syncreticatic nephrotic syndrome, nephritic syndrome, nephrotic-range proteinuria, glomerular disease, and membranous glomerulonephritis.

Effects on the Immune System

Klinghardt's axiom states that "Most, if not all, chronic infectious diseases are not caused by a failure of the immune system, but are a conscious adaptation of the immune system to an otherwise lethal heavy metal environment". It has been known for many years that mercury impairs immune system function most likely via its deleterious effects on the polymorphonuclear leukocytes (PMNs). Mercury through suppression of adrenocorticosteroids production prevents normal stimulation of PMNs production and also affects PMN function by inhibiting their ability to destroy foreign substances. Mercury-sensitive individuals are more likely to have allergies, asthma, and autoimmune-like symptoms, especially rheumatoid-like ones. Mercury can produce an immune response in the central nervous system, induce alterations in immune cell production and function, and modulate the production of interferon gamma and interleukin-2. With impairment comes a chronically susceptible to infections, if not chronic sickness.

Interestingly, the ingestion of mercury is oftentimes associated with increased levels of yeasts, bacteria, and molds which are thought to function in a protective manner to absorb excess mercury from the body. Indiscriminant and rapid destruction of the *Candida albicans* and other pathogens by antibiotics in adults with a significant body burden of toxic metals, including mercury, may cause the sudden release of large amounts of toxic metals contained within them and be potentially very dangerous. Mercury body burden has also been associated with or implicated in a number of immune or autoimmune conditions including allergic disease, amyotrophic lateral sclerosis, arthritis, autoimmune thyroiditis, autism/attention deficit hyperactivity disorder, eczema, epilepsy, psoriasis, multiple sclerosis, rheumatoid arthritis, schizophrenia, scleroderma, and systemic lupus erythematosus.

Effects on the Nervous System

It is clear that mercury is accumulated in nervous tissues all through the body. The most devastating effect of mercury in the nervous system is interference with the production of energy which can impair cellular detoxification processes causing the cell to either die or live in a state of chronic malnutrition. It is thought that mercury causes neuronal problems through blockage of the P-450 enzymatic process. Mercury is associated with increased tissue oxidative damage, and children with autism had significantly higher urinary levels of lipid peroxidation when compared to controls. In the peripheral nervous system, circulating inorganic mercury can be taken up into the nerve terminals where it can impair the synthesis of tubulin and actin which are important constituents of neuronal cell structure and detoxification processes. Primary sensory neuropathy is a hallmark of MeHg poisoning.

In the central nervous system mercury can damage the blood brain barrier and it facilitates penetration of the brain by other toxic metals and substances. The effects of mercury poisoning effects in the central nervous system include depression, paranoia, extreme irritability, hallucinations, an inability to concentrate, memory loss, tremors of the hands, head, lips, tongue, jaw and eyelids, weight loss, perpetually low body temperature, drowsiness, headaches, insomnia, and fatigue. Along with nervous system effects, mercury has also shown to have various effects on other special sensory systems including blindness, retinopathy, optic neuropathy, hearing loss, a reduced sense of smell, and abnormal touch sensation. Autism is a syndrome characterized by impairments in social relatedness, language and communication, a need for routine and sameness, abnormal movements, and sensory dysfunction. Mercury can cause immune, sensory, neurological, motor, and behavioral dysfunctions similar to traits defining or associated with autism leading some to suggest that many cases of autism may be a form of mercury poisoning.

Effects on the Endocrine System

Low exposure levels of mercury may affect the endocrine system in animals and people by disruption of the pituitary, thyroid, adrenal glands and pancreas. It is thought that mercury might impair endocrine function through its ability to reduce hormone-receptor binding or through the inhibition of one or more key enzymes or steps in hormone biosynthesis as is seen in the case of adrenal steroid biosynthesis and the inhibition of 21α-hydroxylase. Hormones that appear to be the most affected by mercury are insulin, estrogen, testosterone, and adrenaline.

Mercury can also inhibit catecholamine degradation through inactivation of S-adenosyl-methionine which can cause the accumulation of epinephrine and hyperhidrosis, tachycardia, ptyalism (hyper salivation) and hypertension. In the adrenal cortex, mercury exposure has been found to be associated with lowered plasma levels of corticosterone. Reduced cortisol production causes a compensatory rise in adrenocorticotropic hormone leading to adrenal hyperplasia. Mercury-induced adrenal hyperplasia may eventually stress the adrenal to a point at which there is adrenal atrophy and may be a causative factor in the development of Addison's disease.

Autopsy studies in 1975 revealed that the thyroid and pituitary retain and accumulate more inorganic mercury than the kidneys. Mercury levels in the pituitary gland ranged from 6.3 to 77 ppb in one study, while another found the mean levels to be 28 ppb, levels found to be neurotoxic and cytotoxic. Low levels of pituitary function are associated with depression and suicidal thoughts, and appear to be a major factor in suicide of teenagers and other vulnerable groups. Because of its effect on the pituitary, mercury is known to cause frequent urination as well as high blood pressure.

The thyroid is one of the largest endocrine glands in the body. The thyroid controls how quickly the body burns energy, makes proteins, and how sensitive the body should be to other hormones. Like the pituitary, the thyroid displays an affinity for accumulating mercury. Mercury blocks thyroid hormone production by occupying iodine-binding sites and inhibiting or altering hormone action leading to the impairment of body temperature control, hypothyroidism, thyroid inflammation and depression.

Like the thyroid, the pancreas is also susceptible to the toxic effects of mercury. Insulin, the molecule involved in diabetes, has three sulfur-binding sites which can be bound by mercury causing the interference with normal biological function and a dysregulation of blood glucose levels.

Effects on the Reproductive System

Mercury can precipitate pathophysiological changes along the hypothalamus-pituitary-adrenal and gonadal axis that may affect reproductive function by altering the circulating of levels of follicle-stimulating hormone (FSH), luteinizing hormone (LH), inhibin, estrogen, progesterone, and the androgens. Reduced fertility among dental assistants with occupational exposure to mercury has been noted. Studies in Hong Kong demonstrated that increased mercury levels were associated with infertility in both men and women. In males, mercury can have adverse effects on spermatogenesis, epididymal sperm count, and testicular weight. Evidence also exists linking mercury with erectile dysfunction. In females, mercury has been shown to inhibit the release of FSH and LH from the anterior pituitary which in turn can effect estrogen and progesterone levels leading to ovarial dysfunction, painful or irregular menstruation, premature menopause, and tipped uterus. There is good evidence linking mercury with menstrual disorders including abnormal bleeding, short, long, irregular cycles, and painful periods.

Fetotoxicity

In addition to reproductive issues, mercury is also associated with the fetotoxicity which can present as miscarriage, spontaneous abortions, stillbirth, and low birth weights. In the neonate, mercury exposure during pregnancy has been linked to neural tube defects, craniofacial malformations, delayed growth, and others. Mercury is known to cross the placenta where it can inhibit fetal brain development resulting in cerebral palsy and psychomotor retardation in the latter stages of development. In primates maternal MeHg blood levels were moderatelyrelated to increased abortion rates and decreased oregnancy rates. Embryopathic effects of MeHg in humans have also been reported. Fetal autopsies indicated a generalized hypoplasia of the cerebellum, decreased number of nerve cells in the cerebral cortex, marked decrease in total brain weight, abnormal neuron migration, and brain centers and layer deranged organization. MeHg easily enters through the placenta and damages the brain of the fetus. Many exposed feti go on to develop infantile cerebral palsy and there may be a relation with the development of Minamata disease. Babies may be born with a variety of birth defects.

Mercury Poisoning

Mercury poisoning is a type of metal poisoning due to exposure to mercury. Symptoms depend upon the type, dose, method, and duration of exposure. They may include muscle weakness, poor coordination, numbness in the hands and feet, skin rashes, anxiety, memory problems, trouble speaking, trouble hearing, or trouble seeing. High-level exposure to methylmercury is known as Minamata disease. Methylmercury exposure in children may result in acrodynia (pink disease) in which the skin becomes pink and peels. Long-term complications may include kidney problems and decreased intelligence. The effects of long-term low-dose exposure to methylmercury are unclear.

Forms of mercury exposure include metal, vapor, salt, and organic compound. Most exposure is from eating fish, amalgam based dental fillings, or exposure at work. In fish, those higher up in the food chain generally have higher levels of mercury. Less commonly, poisoning may occur as a method of attempted suicide. Human activities that release mercury into the environment include the burning of coal and mining of gold. Tests of the blood, urine, and hair for mercury are available but do not relate well to the amount in the body.

Prevention includes eating a diet low in mercury, removing mercury from medical and other devices, proper disposal of mercury, and not mining further mercury. In those with acute poisoning from inorganic mercury salts, chelation with either dimercaptosuccinic acid (DMSA) or dimercaptopropane sulfonate (DMPS) appears to improve outcomes if given within a few hours of exposure. Chelation for those with long-term exposure is of unclear benefit. In certain communities that survive on fishing, rates of mercury poisoning among children have been as high as 1.7 per 100.

Signs and Symptoms

Common symptoms of mercury poisoning include peripheral neuropathy, presenting as paresthesia or itching, burning, pain, or even a sensation that resembles small insects crawling on or under the skin (formication); skin discoloration (pink cheeks, fingertips and toes); swelling; and desquamation (shedding or peeling of skin).

Mercury irreversibly inhibits selenium-dependent enzymes and may also inactivate *S*-adenosyl-methionine, which is necessary for catecholamine catabolism by catechol-*O*-methyl transferase. Due to the body's inability to degrade catecholamines (e.g. epinephrine), a person suffering from mercury poisoning may experience profuse sweating, tachycardia (persistently faster-than-normal heart beat), increased salivation, and hypertension (high blood pressure).

Affected children may show red cheeks, nose and lips, loss of hair, teeth, and nails, transient rashes, hypotonia (muscle weakness), and increased sensitivity to light. Other symptoms may include kidney dysfunction (e.g. Fanconi syndrome) or neuropsychiatric symptoms such as emotional lability, memory impairment, or insomnia.

Thus, the clinical presentation may resemble pheochromocytoma or Kawasaki disease. Desquamation (skin peeling) can occur with severe mercury poisoning acquired by handling elemental mercury.

Causes

The consumption of fish is by far the most significant source of ingestion-related mercury exposure in humans, although plants and livestock also contain mercury due to bioconcentration of mercury from seawater, freshwater, marine and lacustrine sediments, soils, and atmosphere, and due to biomagnification by ingesting other mercury-containing organisms. Exposure to mercury can occur from breathing contaminated air, from eating foods that have acquired mercury residues during processing, from exposure to mercury vapor in mercury amalgam dental restorations, and from improper use or disposal of mercury and mercury-containing objects, for example, after spills of elemental mercury or improper disposal of fluorescent lamps.

All of these, except elemental liquid mercury, produce toxicity or death with less than a gram. Mercury's zero oxidation state (Hg^0) exists as vapor or as liquid metal, its mercurous state (Hg^+) exists as inorganic salts, and its mercuric state (Hg^{2+}) may form either inorganic salts or organomercury compounds.

Consumption of whale and dolphin meat, as is the practice in Japan, is a source of high levels of mercury poisoning.

Human-generated sources, such as coal-burning power plants emit about half of atmospheric mercury, with natural sources such as volcanoes responsible for the remainder. An estimated two-thirds of human-generated mercury comes from stationary combustion, mostly of coal. Other important human-generated sources include gold production, nonferrous metal production, cement production, waste disposal, human crematoria, caustic soda production, pig iron and steel production, mercury production (mostly for batteries), and biomass burning.

Small independent gold-mining operation workers are at higher risk of mercury poisoning because of crude processing methods. Such is the danger for the galamsey in Ghana and similar workers known as *orpailleurs* in neighboring francophone countries. While no official government estimates of the labor force have been made, observers believe 20,000–50,000 work as galamseys in Ghana, a figure including many women, who work as porters. Similar problems have been reported amongst the gold miners of Indonesia.

Some mercury compounds, especially organomercury compounds, can also be readily absorbed through direct skin contact. Mercury and its compounds are commonly used in chemical laboratories, hospitals, dental clinics, and facilities involved in the production of items such as fluorescent light bulbs, batteries, and explosives.

Many traditional medicines, including Ayurvedic medicine and Traditional Chinese medicine contain mercury and other heavy metals.

No scientific data support the claim that mercury compounds in vaccine preservatives cause autism or its symptoms.

Sources

Compounds of mercury tend to be much more toxic than either the elemental form or the salts. These compounds have been implicated in causing brain and liver damage. The most dangerous mercury compound, dimethylmercury, is so toxic that even a few microliters spilled on the skin, or even on a latex glove, can cause death.

Methylmercury and Related Organomercury Compounds

Methylmercury is the major source of organic mercury for all individuals. Due to bioaccumulation it works its way up through the food web and thus biomagnifies, resulting in high concentrations among populations of some species. Top predatory fish, such as tuna or swordfish, are usually of greater concern than smaller species. The US FDA and the EPA advise women of child-bearing age, nursing mothers, and young children to completely avoid swordfish, shark, king mackerel and tilefish from the Gulf of Mexico, and to limit consumption of albacore ("white") tuna to no more than 170 g (6 oz) per week, and of all other fish and shellfish to no more than 340 g (12 oz) per week. A 2006 review of the risks and benefits of fish consumption found, for adults, the benefits of one to two servings of fish per week outweigh the risks, even (except for a few fish species) for women of childbearing age, and that avoidance of fish consumption could result in significant excess coronary heart disease deaths and suboptimal neural development in children.

The period between exposure to methylmercury and the appearance of symptoms in adult poisoning cases is long. The longest recorded latent period is five months after a single exposure, in the Dartmouth case; other latent periods in the range of weeks to months have also been reported. No explanation for this long latent period is known. When the first symptom appears, typically paresthesia (a tingling or numbness in the skin), it is followed rapidly by more severe effects, sometimes ending in coma and death. The toxic damage appears to be determined by the peak value of mercury, not the length of the exposure.

Methylmercury exposure during rodent gestation, a developmental period that approximately models human neural development during the first two trimesters of gestation, has long-lasting behavioral consequences that appear in adulthood and, in some cases, may not appear until aging. Prefrontal cortex or dopamine neurotransmission could be especially sensitive to even subtle gestational methylmercury exposure and suggests that public health assessments of methylmercury based on intellectual performance may underestimate the impact of methylmercury in public health.

Ethylmercury is a breakdown product of the antibacteriological agent ethylmercurithiosalicylate, which has been used as a topical antiseptic and a vaccine preservative. Its characteristics have not been studied as extensively as those of methylmercury. It is cleared from the blood much more rapidly, with a half-life of seven to ten days, and it is metabolized much more quickly than methylmercury. It is presumed not to have methylmercury's ability to cross the blood–brain barrier via a transporter, but instead relies on simple diffusion to enter the brain. Other exposure sources of organic mercury include phenylmercuric acetate and phenylmercuric nitrate. These compounds were used in indoor latex paints for their antimildew properties, but were removed in 1990 because of cases of toxicity.

Inorganic Mercury Compounds

Mercury occurs as salts such as mercuric chloride ($HgCl_2$) and mercurous chloride (Hg_2Cl_2), the latter also known as calomel. Because they are more soluble in water, mercuric salts are usually more acutely toxic than mercurous salts. Their higher solubility lets them be more readily absorbed from the gastrointestinal tract. Mercury salts affect primarily the gastrointestinal tract and the kidneys, and can cause severe kidney damage; however, as they cannot cross the blood–brain barrier easily, these salts inflict little neurological damage without continuous or heavy exposure. Mercuric cyanide

($Hg(CN)_2$) is a particularly toxic mercury compound that has been used in murders, as it contains not only mercury but also cyanide, leading to simultaneous cyanide poisoning. The drug n-acetyl penicillamine has been used to treat mercury poisoning with limited success.

Elemental Mercury

Quicksilver (liquid metallic mercury) is poorly absorbed by ingestion and skin contact. Its vapor is the most hazardous form. Animal data indicate less than 0.01% of ingested mercury is absorbed through the intact gastrointestinal tract, though it may not be true for individuals suffering from ileus. Cases of systemic toxicity from accidental swallowing are rare, and attempted suicide via intravenous injection does not appear to result in systemic toxicity, though it still causes damage by physically blocking blood vessels both at the site of injection and the lungs. Though not studied quantitatively, the physical properties of liquid elemental mercury limit its absorption through intact skin and in light of its very low absorption rate from the gastrointestinal tract, skin absorption would not be high. Some mercury vapor is absorbed dermally, but uptake by this route is only about 1% of that by inhalation.

In humans, approximately 80% of inhaled mercury vapor is absorbed via the respiratory tract, where it enters the circulatory system and is distributed throughout the body. Chronic exposure by inhalation, even at low concentrations in the range 0.7–42 μg/m³, has been shown in case–control studies to cause effects such as tremors, impaired cognitive skills, and sleep disturbance in workers.

Acute inhalation of high concentrations causes a wide variety of cognitive, personality, sensory, and motor disturbances. The most prominent symptoms include tremors (initially affecting the hands and sometimes spreading to other parts of the body), emotional lability (characterized by irritability, excessive shyness, confidence loss, and nervousness), insomnia, memory loss, neuromuscular changes (weakness, muscle atrophy, muscle twitching), headaches, polyneuropathy (paresthesia, stocking-glove sensory loss, hyperactive tendon reflexes, slowed sensory and motor nerve conduction velocities), and performance deficits in tests of cognitive function.

Mechanism

The toxicity of mercury sources can be expected to depend on its nature, i.e., salts vs. organomercury compounds vs. elemental mercury.

One mechanism of mercury toxicity involves its irreversible inhibition of selenoenzymes, such as thioredoxin reductase (IC50 = 9 nM). Although it has many functions, thioredoxin reductase restores vitamins C and E, as well as a number of other important antioxidant molecules, back into their reduced forms, enabling them to counteract oxidative damage. Since the rate of oxygen consumption is particularly high in brain tissues, production of reactive oxygen species (ROS) is accentuated in these vital cells, making them particularly vulnerable to oxidative damage and especially dependent upon the antioxidant protection provided by selenoenzymes. High mercury exposures deplete the amount of cellular selenium available for the biosynthesis of thioredoxin reductase and other selenoenzymes that prevent and reverse oxidative damage, which, if the depletion is severe and long lasting, results in brain cell dysfunctions that can ultimately cause death.

Mercury in its various forms is particularly harmful to fetuses as an environmental toxin in pregnancy, as well as to infants. Women who have been exposed to mercury in substantial excess of dietary selenium intakes during pregnancy are at risk of giving birth to children with serious birth defects. Mercury exposures in excess of dietary selenium intakes in young children can have severe neurological consequences, preventing nerve sheaths from forming properly.

Exposure to methylmercury causes increased levels of antibodies sent to myelin basic protein (MBP), which is involved in the myelination of neurons, and glial fibrillary acidic protein (GFAP), which is essential to many central nervous system (CNS). This causes an autoimmmune response against MBP and GFAP and results in the degradation of neural myelin and general decline in function of the CNS.

Diagnosis

Diagnosis of elemental or inorganic mercury poisoning involves determining the history of exposure, physical findings, and an elevated body burden of mercury. Although whole-blood mercury concentrations are typically less than 6 μg/L, diets rich in fish can result in blood mercury concentrations higher than 200 μg/L, it is not that useful to measure these levels for suspected cases of elemental or inorganic poisoning because of mercury's short half-life in the blood. If the exposure is chronic, urine levels can be obtained; 24-hour collections are more reliable than spot collections. It is difficult or impossible to interpret urine samples of patients undergoing chelation therapy, as the therapy itself increases mercury levels in the samples.

Diagnosis of organic mercury poisoning differs in that whole-blood or hair analysis is more reliable than urinary mercury levels.

Chromium Toxicity

Chromium toxicity refers to any poisonous toxic effect in an organism or cell that results from exposure to specific forms of chromium—especially hexavalent chromium. Hexavalent chromium and its compounds are toxic when inhaled or ingested. Trivalent chromium is a trace mineral that is essential to human nutrition. There is a hypothetical risk of genotoxicity in humans if large amounts of trivalent chromium were somehow able to enter living cells, but normal metabolism and cell function prevent this.

Forms of Chromium

Hexavalent chromium and trivalent chromium are chromium ions—they have different numbers of electrons and, therefore, different properties. Trivalent chromium, or chromium(III), is the form of chromium that is essential to human health. Hexavalent chromium, or chromium(VI), is an unequivocally toxic form.

Hexavalent Chromium

Hexavalent chromium, also called chromium(VI), is hemotoxic, genotoxic, and carcinogenic. When hexavalent chromium enters the bloodstream, it damages blood cells by causing oxidation reactions. This oxidative damage can lead to hemolysis and, ultimately, kidney and liver failure. Patients might be treated with dialysis.

The median lethal dose of hexavalent chromium is 50–150 mg/kg. The World Health Organization recommends a maximum allowable concentration of 0.05 milligrams per litre of chromium(VI) in drinking water. In Europe, the use of hexavalent chromium is regulated by the Restriction of Hazardous Substances Directive.

Hexavalent chromium can be found in some dyes and paints, as well as in some leather tanning products. Primer paint containing hexavalent chromium is widely used in aerospace and automobile refinishing applications. Metal workers (such as welders)—as well as people with a surgical implant made from cobalt-chromium alloy—may also be exposed to hexavalent chromium. Chromium concentrations in whole blood, plasma, serum, or urine may be measured to monitor for safety in exposed workers, to confirm the diagnosis in potential poisoning victims, or to assist in the forensic investigation in a case of fatal overdosage.

In the U.S. state of California, an epidemic of hexavalent chromium exposure led to a class-action lawsuit in 1993: *Anderson, v. Pacific Gas and Electric*. The Pacific Gas and Electric Company had dumped more than 1.4 billion litres (370 million gallons) of wastewater tainted with hexavalent chromium into the Mojave Desert. This contaminated the groundwater, and caused widespread illness among the people of Hinkley, California, a small community nearby. As of May 2017, the mandated environmental remediation measures are ongoing.

Chromate

Chromates (chromium salts) formed from hexavalent chromium are used to manufacture leather products, paints, cement, mortar, anti-corrosives, and other things. They are carcinogenic and allergenic. The carcinogenity of chromate dust has been documented since the late 19th century, when workers in a chromate dye company were found to exhibit high incidence of cancer. Chromate enters cells by means of the same transport mechanism that carries sulfate and phosphate ions into cells.

Contact with products containing chromates can lead to allergic contact dermatitis and irritant dermatitis, resulting in ulceration of the skin—a condition sometimes called *chrome ulcers*. Workers that have been exposed to strong chromate solutions in electroplating, tanning, and chrome-producing manufacturers may also develop chrome ulcers.

Genotoxicity

Hexavalent chromium is genotoxic: it damages genetic information in living cells, which results in DNA mutations, and possibly the formation of cancerous tumors. One hypothesis asserts that the genotoxicity is caused by free radicals such as hydroxyl radicals, produced by the reduction of chromium(VI) to chromium(III). Another proposed mechanism supposes that chromium binds to DNA at the end of the reduction to chromium(III).

A third hypothesis proposes that two other forms of chromium—Chromium(IV) compounds, along with chromium(V) produced by redox reactions in the cell—bind to DNA.

Trivalent Chromium

Trivalent chromium, or chromium(III), is an essential trace mineral in the human diet. In some nutritional supplements, chromium(III) occurs as chromium(III) picolinate (in which chromium is bound to picolinic acid) or chromium(III) nicotinate (in which chromium is bound to nicotinic acid). Nicotinic acid is also known as the B vitamin niacin.

Chromium(III) is poorly absorbed in humans; most dietary chromium is excreted in the urine. The threshold for acute oral toxicity is 1900–3300 mg/kg. In rats, nonsteroidal anti-inflammatory drugs such as aspirin and indometacin can increase chromium absorption.

Ordinarily, cellular transport mechanisms in humans and some other animals limit the amount of chromium(III) that enters a cell. Hypothetically, if an excessive amount was able to enter a cell, free radical damage to DNA might result.

Benzene

Human exposure to benzene has been associated with a range of acute and long-term adverse health effects and diseases, including cancer and aplastic anaemia. Exposure can occur occupationally and domestically as a result of the ubiquitous use of benzene-containing petroleum products, including motor fuels and solvents. Active and passive exposure to tobacco smoke is also a significant source of exposure. Benzene is highly volatile, and exposure occurs mostly through inhalation. Public health actions are needed to reduce the exposure of both workers and the general population to benzene.

Sources of Exposure to Benzene

Benzene is highly volatile, and most exposure is through inhalation. Benzene is degraded rapidly in the upper atmosphere. Because of its solubility in water, a minor amount may be removed by rain to contaminate surface waters and soil. However, it is not persistent in surface water or soil, either volatilizing back to air or being degraded by bacteria.

Industrial Processes

As benzene occurs naturally in crude petroleum at levels up to 4 g/l, human activities using petroleum lead to exposure. These activities include processing of petroleum products, coking of coal, production of toluene, xylene and other aromatic compounds, and use in industrial and consumer products, as a chemical intermediate and as a component of petrol (gasoline) and heating oils. The presence of benzene in petrol and as a widely used industrial solvent can result in significant occupational exposure and widespread emissions to the environment. Automobile exhaust accounts for the largest source of benzene in the general environment. Off-gassing from building materials and structural fires lead to increased atmospheric benzene levels. Industrial discharge, landfill leachate and disposal of benzenecontaining waste are also sources of exposure.

Indoor Residential Air

Benzene has been detected at high levels in indoor air. Although some of this exposure might be from building materials (paints, adhesives, etc.), most is from cigarette smoke in both homes and public spaces. Levels of benzene are higher in homes with attached garages than in those with detached garages. Levels are increased in homes close to petrol filling stations. Benzene may be released to indoor air from unflued oil heating and from the use of benzenecontaining consumer products in residences. People spending more time indoors, such as children, are likely to have higher exposure to benzene.

Inside Vehicles

Benzene has been measured in air inside vehicles at levels higher than those in residential air, but substantially lower than those at petrol filling stations.

Food and Water

Waterborne and foodborne benzene contributes only a small percentage of the total daily intake in non-smoking adults.

Health Effects

Acute effects:

- Acute occupational exposure to benzene may cause narcosis: headache, dizziness, drowsiness, confusion, tremors and loss of consciousness. Use of alcohol enhances the toxic effect.
- Benzene is a moderate eye irritant and a skin irritant.

Effects following chronic exposure:

- Benzene is a well-established cause of cancer in humans. The International Agency for Research on Cancer has classified benzene as carcinogenic to humans. Benzene causes acute myeloid leukaemia (acute non-lymphocytic leukaemia), and there is limited evidence that benzene may also cause acute and chronic lymphocytic leukaemia, non-Hodgkin's lymphoma and multiple myeloma. Individuals who have experienced benzene poisoning requiring treatment show a substantially increased risk of mortality from leukaemia.
- Chronic exposure to benzene can reduce the production of both red and white blood cells from bone marrow in humans, resulting in aplastic anaemia.
- Both B-cell proliferation and T-cell proliferation are reduced by benzene. Decreased host resistance to infection has been reported in several laboratory animals exposed to benzene. However, other measures of immunotoxicity have not been studied.
- Chromosomal aberrations in human peripheral lymphocytes are associated with occupational exposure to benzene. Chromosomal aberrations, micronuclei, sister chromatid exchange and sperm head abnormalities have been seen in laboratory species treated in vivo. Chromosomal aberrations and mutations were seen in human cells in vitro and in laboratory animal cells in some in vitro studies.
- Benzene is fetotoxic in mice and rabbits following maternal exposure by inhalation, causing a reduction in birth weight.2 It is not, however, teratogenic in experimental animals, even at maternally toxic doses.

Monitoring Exposure

- Estimates of benzene exposure can be made through measurement of urinary inorganic and organic sulfate conjugates. Measurement of urinary phenol levels has historically been the standard bioassay for benzene exposure, despite the welldescribed limitations of this test.
- The haematological effects of chronic benzene poisoning in exposed workers can be detected by monitoring blood counts at regular intervals. For example,

the United States Occupational Safety and Health Administration recommends monthly blood counts and removal of workers from areas with high benzene exposure if they have white blood cell counts below 4000/mm3 or erythrocyte counts below 4 000 000/mm.

Risk Mitigation Recommendations

Eliminate Use

- Promote the use of alternative solvents in industrial processes, glues and paints.
- Develop and implement policies and legislation to remove benzene from consumer products.

Reduce Exposure

- Reduce exposure at petrol filling stations as far as possible by following best practices in location, design and extraction.
- Minimize emissions from vehicle exhausts by improved design and regular monitoring of engine settings.
- Separate dwelling spaces from areas where vehicles and benzene-containing products are kept. In particular, isolate children from indoor exposure to vehicle emissions.
- Avoid domestic use of benzene-containing products.
- Discourage indoor use of unflued oil and gasoline heating.
- Prohibit smoking inside buildings.

Formaldehyde

Formaldehyde is used mainly to produce resins used in particleboard products and as an intermediate in the synthesis of other chemicals. Exposure to formaldehyde may occur by breathing contaminated indoor air, tobacco smoke, or ambient urban air. Acute (short-term) and chronic (long-term) inhalation exposure to formaldehyde in humans can result in respiratory symptoms, and eye, nose, and throat irritation. Limited human studies have reported an association between formaldehyde exposure and lung and nasopharyngeal cancer. Animal inhalation studies have reported an increased incidence of nasal squamous cell cancer.

The main sources of information for this fact sheet are EPA's Health and Environmental Effects Profile for Formaldehyde and the Integrated Risk Information System (IRIS), which contains information on oral chronic toxicity and the RfD, and the carcinogenic effects of formaldehyde including the unit cancer risk for inhalation exposure.

Uses

- Formaldehyde is used predominantly as a chemical intermediate. It also has minor uses in agriculture, as an analytical reagent, in concrete and plaster additives, cosmetics, disinfectants, fumigants, photography, and wood preservation.
- One of the most common uses of formaldehyde is manufacturing urea-formaldehyde resins, used in particleboard products.
- Formaldehyde (as urea formaldehyde foam) was extensively used as an insulating material until 1982 when it was banned.

Sources and Potential Exposure

- The highest levels of airborne formaldehyde have been detected in indoor air, where it is released from various consumer products such as building materials and home furnishings. One survey reported formaldehyde levels ranging from 0.10 to 3.68 parts per million (ppm) in homes. Higher levels have been found in new manufactured or mobile homes than in older conventional homes.
- Formaldehyde has also been detected in ambient air; the average concentrations reported in U.S. urban areas were in the range of 11 to 20 parts per billion (ppb).
- The major sources appear to be power plants, manufacturing facilities, incinerators, and automobile exhaust emissions.
- Smoking is another important source of formaldehyde.
- Formaldehyde may also be present in food, either naturally or as a result of contamination.

Assessing Personal Exposure

- Blood levels of formaldehyde can be measured. However, these measurements are only useful when exposure to relatively large amounts of formaldehyde has occurred.

Health Hazard Information

Acute Effects:

- Acute Effects: The major toxic effects caused by acute formaldehyde exposure via inhalation are eye, nose, and throat irritation and effects on the nasal cavity. Other effects seen from exposure to high levels of formaldehyde in humans are coughing, wheezing, chest pains, and bronchitis.

- Ingestion exposure to formaldehyde in humans has resulted in corrosion of the gastrointestinal tract and inflammation and ulceration of the mouth, esophagus, and stomach.
- Acute animal tests in rats and rabbits have shown formaldehyde to have high acute toxicity from inhalation, oral, and dermal exposure.

Chronic Effects (Noncancer)

- Chronic exposure to formaldehyde by inhalation in humans has been associated with respiratory symptoms and eye, nose, and throat irritation.
- Repeated contact with liquid solutions of formaldehyde has resulted in skin irritation and allergic contact dermatitis in humans.
- Animal studies have reported effects on the nasal respiratory epithelium and lesions in the respiratory system from chronic inhalation exposure to formaldehyde.
- The Reference Dose (RfD) for formaldehyde is 0.2 milligrams per kilogram body weight per day (mg/kg/d) based on a decrease in body weight gain and effects on the stomach in rats. The RfD is an estimate (with uncertainty spanning perhaps an order of magnitude) of a daily oral exposure to the human population (including sensitive subgroups) that is likely to be without appreciable risk of deleterious noncancer effects during a lifetime. It is not a direct estimator of risk but rather a reference point to gauge the potential effects. At exposures increasingly greater than the RfD, the potential for adverse health effects increases. Lifetime exposure above the RfD does not imply that an adverse health effect would necessarily occur.
- EPA has high confidence in the study on which the RfD was based since it consisted of an adequate number of animals of both sexes, as well as a thorough examination of toxicological and histological parameters; medium confidence in the database as several additional chronic bioassays and reproductive and developmental studies support the critical effect and study; and, consequently, medium confidence in the RfD.
- EPA has not established a Reference Concentration (RfC) for formaldehyde.
- The Agency for Toxic Substances and Disease Registry (ATSDR) has established a chronic inhalation 3 minimal risk level (MRL) of 0.003 ppm (0.004 milligrams per cubic meter, mg/m) based on respiratory effects in humans. The MRL is an estimate of the daily human exposure to a hazardous substance that is likely to be without appreciable risk of adverse noncancer health effects over a specified duration of exposure.

Reproductive/Developmental Effects

- An increased incidence of menstrual disorders were observed in female workers using urea-formaldehyde resins. However, possible confounding factors were not evaluated in this study.
- A study of hospital equipment sterilizing workers did not report an association between formaldehyde exposure and increased spontaneous abortions.
- Developmental effects, such as birth defects, have not been observed in animal studies with formaldehyde.

Cancer Risk

- Occupational studies have noted statistically significant associations between exposure to formaldehyde and increased incidence of lung and nasopharyngeal cancer. This evidence is considered to be "limited," rather than "sufficient," due to possible exposure to other agents that may have contributed to the excess cancers.
- Animal studies have reported an increased incidence of nasal squamous cell carcinomas by inhalation exposure.
- EPA considers formaldehyde to be a probable human carcinogen (cancer-causing agent) and has ranked it in EPA's Group B1.
- EPA uses mathematical models, based on animal studies, to estimate the probability of a person developing cancer from breathing air containing a specified concentration of a chemical. EPA calculated an inhalation -5 3 -1 unit risk estimate of 1.3×10^{-5} ($\mu g/m^3$). EPA estimates that, if an individual were to continuously breathe 3 -5 3 air containing formaldehyde at an average of 0.08 µg/m (8.0 x 10 mg/m) over his or her entire lifetime, that person would theoretically have no more than a one-in-a-million increased chance of developing cancer as a direct result of breathing air containing this chemical. Similarly, EPA estimates that 3 -4 3 breathing air containing 0.8 $\mu g/m^3$ (8.0 x 10 mg/m) would result in not greater than a one-in-a3 hundred thousand increased chance of developing cancer, and air containing 8.0 $\mu g/m^3$ (8.0×10^{-4} mg/m^3) would result in not greater than a one-in-ten-thousand increased chance of developing cancer.

Radon

Radon is colorless at standard temperature and pressure and it is the most dense gas known. At temperature below it's freezing point is has a brilliant yellow phosphorescence. It is chemically unreactive, it is highly radioactive and has a short half life.

Applications

Radon was sometimes used in hospitals to treat cancer and was produced as needed and delivered in sealed gold needles. Radon is used in hydrologic research, because of it's rapid loss to air. It is also used in geologic research and to track air masses.

Radon in the Environment

Radon can be found in some spring water and hot springs. There is anyway a detectable amount of radon in the atmosphere. Radon collects over samples of radium 226 at the rate of around 0.001 cm3/day per g of radium.

Health Effects of Radon

Radon occurs in the environment mainly in the gaseous phase. Consequently, people are mainly exposed to radon through breathing air.

Background levels of radon in outside air are generally quite low, but in indoor locations radon levels in air may be higher. In homes, schools and buildings radon levels are increased because radon enters the buildings through cracks in the foundations and basements.

Some of the deep wells that supply us with drinking water may also contain radon. As a result a number of people may be exposed to radon through drinking water, as well as through breathing air.

Radon levels in groundwater are fairly high, but usually radon is quickly released into air as soon as the groundwater enters surface waters.

Exposure to high levels of radon through breathing air is known to cause lung diseases. When long-term exposure occurs radon increases the chances of developing lung cancer. Radon can only cause cancer after several years of exposure.

Radon may be radioactive, but it gives off little actual gamma radiation. As a result, harmful effects from exposure to radon radiation without actual contact with radon compounds are not likely to occur.

It is not known whether radon can cause health effects in other organs besides the lungs. The effects of radon, which is found in food or drinking water, are unknown.

Environmental Effects of Radon

Radon is a radioactive compound, which rarely occurs naturally in the environment. Most of the radon compounds found in the environment derive from human activities. Radon enters the environment through the soil, through uranium and phosphate mines, and through coal combustion.

Some of the radon that is located in the soil will move to the surface and enter the air through vaporization. In the air, radon compounds will attach to dust and other particles. Radon can also move downwards in the soil and enter the groundwater. However, most of the radon will remain in the soil.

Radon has a radioactive half-life of about four days; this means that one-half of a given amount of radon will decay to other compounds, usually less harmful compounds, every four days.

Cyanotoxins

Cyanotoxins are toxins produced by bacteria called cyanobacteria (also known as blue-green algae). Cyanobacteria are found almost everywhere, but particularly in lakes and in the ocean where, under high concentration of phosphorus conditions, they reproduce exponentially to form blooms. Blooming cyanobacteria can produce cyanotoxins in such concentrations that they poison and even kill animals and humans. Cyanotoxins can also accumulate in other animals such as fish and shellfish, and cause poisonings such as shellfish poisoning.

Green scum produced by and containing cyanobacteria, washed up on a rock in california during an algal bloom.

Among cyanotoxins are some of the most powerful natural poisons known, including poisons which can cause rapid death by respiratory failure. The toxins include potent neurotoxins, hepatotoxins, cytotoxins, and endotoxins. Despite the similarity in name, they are not cyanides. Recreational exposure to cyanobacteria can result in gastro-intestinal and hay fever symptoms or pruritic skin rashes. Exposure to the cyanobacteria neurotoxin BMAA may be an environmental cause of neurodegenerative diseases such as amyotrophic lateral sclerosis (ALS), Parkinson's disease, and Alzheimer's disease. There is also an interest in the military potential of biological neurotoxins such as cyanotoxins, which "have gained increasing significance as potential candidates for weaponization."

The first published report that blue-green algae or cyanobacteria could have lethal effects appeared in *Nature* in 1878. George Francis described the algal bloom he observed in the estuary of the Murray River in Australia, as "a thick scum like green oil paint, some two to six inches thick." Wildlife which drank the water died rapidly and terribly. Most reported incidents of poisoning by microalgal toxins have occurred in freshwater environments, and they are becoming more common and widespread. For example, thousands of ducks and geese died drinking contaminated water in the midwestern United States. In 2010, for the first time, marine mammals were reported to have died from ingesting cyanotoxins.

Cyanobacteria

Cyanotoxins are produced by cyanobacteria, a phylum of bacteria that obtain their energy through photosynthesis. The prefix *cyan* comes from the Greek κύανος meaning «a dark blue substance», and usually indicates any of a number of colours in the blue/green range of the spectrum. Cyanobacteria are commonly referred to as *blue-green algae*. Traditionally they were thought of as a form of algae, and were introduced as such in older textbooks. However modern sources tend to regard this as outdated; they are now considered to be more closely related to bacteria, and the term for true *algae* is restricted to eukaryotic organisms. Like true algae, cyanobacteria are photosynthetic and contain photosynthetic pigments, which is why they are usually green or blue.

Cyanobacteria are found almost everywhere; in oceans, lakes and rivers as well as on land. They flourish in Arctic and Antarctic lakes, hotsprings and wastewater treatment plants. They even inhabit the fur of polar bears, to which they impart a greenish tinge. Cyanobacteria produce potent toxins, but they also produce helpful bioactive compounds, including substances with antitumour, antiviral, anticancer, antibiotic and antifungal activity, UV protectants and specific inhibitors of enzymes.

Harmful Algal Blooms

Cyanotoxins are often implicated in what are commonly called *red tides* or *harmful algal blooms*. Lakes and oceans contain many single-celled organisms called phytoplankton. Under certain conditions, particularly when nutrient concentrations are high, these organisms reproduce exponentially. The resulting dense swarm of phytoplankton is called an algal bloom; these can cover hundreds of square kilometres and can be easily seen in satellite images. Individual phytoplankton rarely live more than a few days, but blooms can last weeks.

Generally, these blooms are harmless, but if not they are called harmful algal blooms, or HABs. HABs can contain toxins or pathogens which result in fish kill and can also be fatal to humans. In marine environments, HABs are mostly caused by dinoflagellates, though species of other algae taxa can also cause HABs (diatoms, flagellates, haptophytes and raphidophytes). Marine dinoflagellate species are often toxic, but freshwater species are not known to be toxic. Neither are diatoms known to be toxic, at least to humans.

Dense bloom of cyanobacteria on the Potomac River estuary. These blooms can be toxic.

In freshwater ecosystems, algal blooms are most commonly caused by high levels of nutrients (eutrophication). The blooms can look like foam, scum or mats or like paint floating on the surface of the water, but they are not always visible. Nor are the blooms always green; they can be blue, and some cyanobacteria species are coloured brownish-red. The water can smell bad when the cyanobacteria in the bloom die.

Strong cyanobacterial blooms reduce visibility to one or two centimetres. Species which are not reliant on sight (such as cyanobacteria themselves) survive, but species which need to see to find food and partners are compromised. During the day blooming cyanobacteria saturate the water with oxygen. At night respiring aquatic organisms can deplete the oxygen to the point where sensitive species, such as certain fish, die. This is more likely to happen near the sea floor or a thermocline. Water acidity also cycles daily during a bloom, with the pH reaching 9 or more during the day and dropping to low values at night, further stressing the ecosystem. In addition, many cyanobacteria species produce potent cyanotoxins which concentrate during a bloom to the point where they become lethal to nearby aquatic organisms and any other animals in direct contact with the bloom, including birds, livestock, domestic animals and sometimes humans.

In 1991 a harmful cyanobacterial bloom affected 1,000 km of the Darling-Barwon River in Australia at an economic cost of $10M AUD.

Chemical Structure

Cyanotoxins usually target the nervous system (neurotoxins), the liver (hepatotoxins) or the skin (dermatoxins). The chemical structure of cyanotoxins falls into three broad groups: cyclic peptides, alkaloids and lipopolysaccharides (endotoxins).

Table: Chemical structure of cyanotoxins.

Structure	Cyanotoxin	Primary target organ in mammals
Cyclic peptides	Microcystins	Liver
	Nodularins	Liver
Alkaloids	Anatoxin-a	Nerve synapse
	Anatoxin-a(S)	Nerve synapse
	Cylindrospermopsins	Liver
	Lyngbyatoxin-a	Skin, gastro-intestinal tract
	Saxitoxin	Nerve synapse
	Lipopolysaccharides	Potential irritant; affects any exposed tissue
Polyketides	Aplysiatoxins	Skin
Amino Acid	BMAA	Nervous System

Most cyanotoxins have a number of variants. As of 1999, altogether over 84 cyanotoxins were known and only a small number have been well studied.

Cyclic Peptides

A peptide is a short polymer of amino acids linked by peptide bonds. They have the same chemical structure as proteins, except they are shorter. In a cyclic peptide, the ends link to form a stable circular chain. In mammals this stability makes them resistant to the process of digestion and they can bioaccumulate in the liver. Of all the cyanotoxins, the cyclic peptides are of most concern to human health. The microcystins and nodularins poison the liver, and exposure to high doses can cause death. Exposure to low doses in drinking water over a long period of time may promote liver and other tumours.

Microcystins

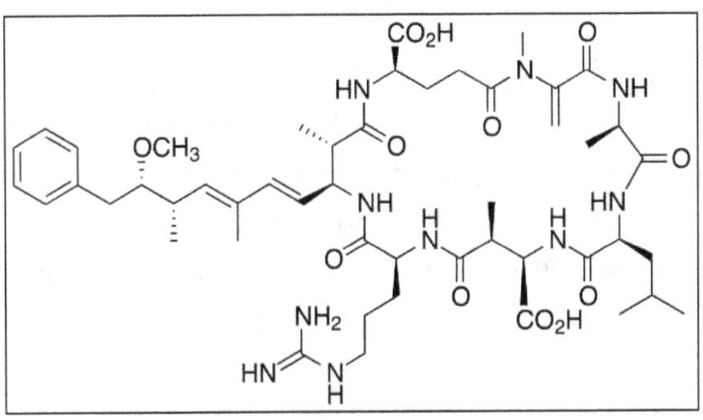

Microcystin LR.

As with other cyanotoxins, microcystins were named after the first organism discovered to produce them, *Microcystis aeruginosa*. However it was later found other cyanobacterial genera also produced them. There are about 60 known variants of microcystin, and several of these can be produced during a bloom. The most reported variant is microcystin-LR, possibly because the earliest commercially available chemical standard analysis was for microcystin-*LR*.

Blooms containing microcystin are a problem worldwide in freshwater ecosystems. Microcystins are cyclic peptides and can be very toxic for plants and animals including humans. They bioaccumulate in the liver of fish, in the hepatopancreas of mussels, and in zooplankton. They are hepatotoxic and can cause serious damage to the liver in humans. In this way they are similar to the nodularins , and together the microcystins and nodularins account for most of the toxic cyanobacterial blooms in fresh and brackish waters. In 2010, a number of sea otters were poisoned by microcystin. Marine bivalves were the likely source of hepatotoxic shellfish poisoning. This was the first confirmed example of a marine mammal dying from ingesting a cyanotoxin.

Nodularins

The first nodularin variant to be identified was nodularin-R, produced by the cyanobacterium *Nodularia spumigena*. This cyanobacterium blooms in water bodies throughout the world. In the Baltic Sea, marine blooms of *Nodularia spumigena* are among some of the largest cyanobacterial mass events in the world. Parts of nine industrialized countries drain into the Baltic Sea, which has little water exchange with the North Sea and Atlantic Ocean. It is consequently one of the more polluted bodies of water in the world (nutrient-rich, from the perspective of cyanobacteria).

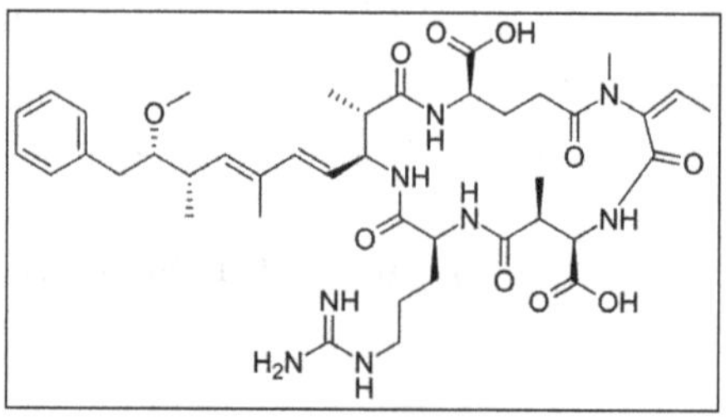

Nodularin-R.

Globally, the most common toxins present in cyanobacterial blooms in fresh and brackish waters are the cyclic peptide toxins of the nodularin family. Like the microcystin family, nodularins are potent hepatotoxins and can cause serious damage to the liver. They present health risks for wild and domestic animals as well as humans, and in many areas pose major challenges for the provision of safe drinking water.

Alkaloids

Alkaloids are a group of naturally occurring chemical compounds which mostly contain basic nitrogen atoms. They are produced by a large variety of organisms, including cyanobacteria, and are part of the group of natural products, also called secondary metabolites. Alkaloids act on diverse metabolic systems in humans and other animals, often with psychotropic or toxic effects. Almost uniformly, they are bitter tasting.

Anatoxin-*a*

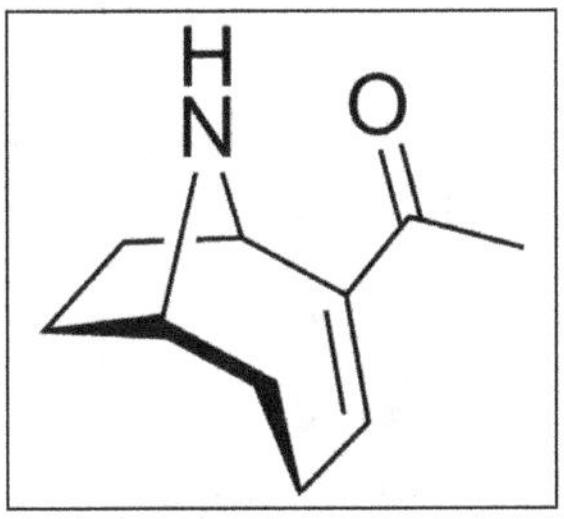

Anatoxin-*a*.

Investigations into anatoxin-*a*, also known as "Very Fast Death Factor", began in 1961 following the deaths of cows that drank from a lake containing an algal bloom in Saskatchewan, Canada. The toxin is produced by at least four different genera of cyanobacteria and has been reported in North America, Europe, Africa, Asia, and New Zealand.

Toxic effects from anatoxin-*a* progress very rapidly because it acts directly on the nerve cells (neurons) as a neurotoxin. The progressive symptoms of anatoxin-*a* exposure are loss of coordination, twitching, convulsions and rapid death by respiratory paralysis. The nerve tissues which communicate with muscles contain a receptor called the nicotinic acetylcholine receptor. Stimulation of these receptors causes a muscular contraction. The anatoxin-*a* molecule is shaped so it fits this receptor, and in this way it mimics the natural neurotransmitter normally used by the receptor, acetylcholine. Once it has triggered a contraction, anatoxin-*a* does not allow the neurons to return to their resting state, because it is not degraded by cholinesterase which normally performs this function. As a result, the muscle cells contract permanently, the communication between the brain and the muscles is disrupted and breathing stops.

The toxin was called the Very Fast Death Factor because it induced tremors, paralysis and death within a few minutes when injected into the body cavity of mice. In 1977, the structure of VFDF was determined as a secondary, bicyclic amine alkaloid, and it was renamed anatoxin-*a*. Structurally, it is similar to cocaine. There is continued interest in anatoxin-*a* because of the dangers it presents to recreational and drinking waters, and because it is a particularly useful molecule for investigating acetylcholine receptors in the nervous system. The deadliness of the toxin means that it has a high military potential as a toxin weapon.

Cylindrospermopsins

Cylindrospermopsin (abbreviated to CYN or CYL) was first discovered after an outbreak of a mystery disease on Palm Island in Australia. The outbreak was traced back to a bloom of *Cylindrospermopsis raciborskii* in the local drinking water supply, and the toxin was subsequently identified. Analysis of the toxin led to a proposed chemical structure in 1992, which was revised after synthesis was achieved in 2000. Several variants of cylindrospermopsin, both toxic and non-toxic, have been isolated or synthesised.

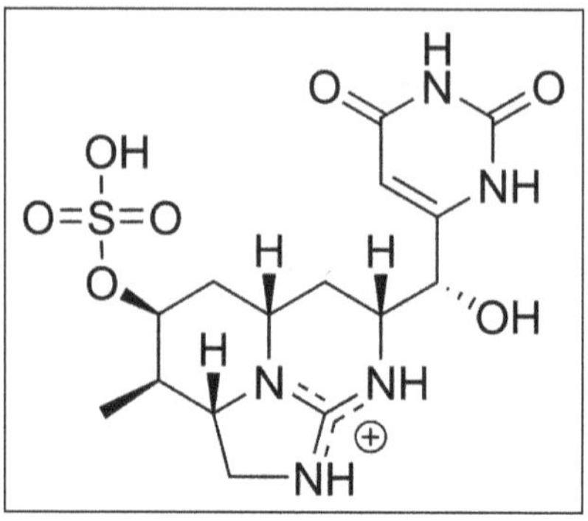

Cylindrospermopsin.

Cylindrospermopsin is toxic to liver and kidney tissue and is thought to inhibit protein synthesis and to covalently modify DNA and RNA. There is concern about the way cylindrospermopsin bioaccumulates in freshwater organisms. Toxic blooms of genera which produce cylindrospermopsin are most commonly found in tropical, subtropical and arid zone water bodies, and have recently been found in Australia, Europe, Israel, Japan and the USA.

Saxitoxins

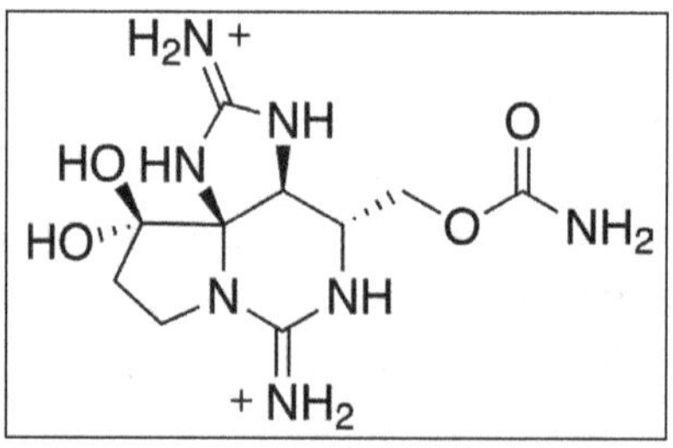

Saxitoxin.

Saxitoxin (STX) is one of the most potent natural neurotoxins known. The term saxitoxin originates from the species name of the butter clam (*Saxidomus giganteus*) whereby it was first recognized. Saxitoxin is produced by the cyanobacteria *Anabaena* spp., some *Aphanizomenon* spp., *Cylindrospermopsis* sp., *Lyngbya* sp. and *Planktothrix* sp., among others). Puffer fish and some marine dinoflagellates also produce saxitoxin. Saxitoxins bioaccumulate in shellfish and certain finfish. Ingestion of saxitoxin, usually through shellfish contaminated by toxic algal blooms, can result in paralytic shellfish poisoning.

Saxitoxin has been used in molecular biology to establish the function of the sodium channel. It acts on the voltage-gated sodium channels of nerve cells, preventing normal cellular function and leading to paralysis. The blocking of neuronal sodium channels which occurs in paralytic shellfish poisoning produces a flaccid paralysis that leaves its victim calm and conscious through the progression of symptoms. Death often occurs from respiratory failure. Saxitoxin was originally isolated and described by the United

States military, who assigned it the chemical weapon designation "TZ". Saxitoxin is listed in schedule 1 of the Chemical Weapons Convention.

Lipopolysaccharides

Lipopolysaccharides are present in all cyanobacteria. Though not as potent as other cyanotoxins, some researchers have claimed that all lipopolysaccharides in cyanobacteria can irritate the skin, while other researchers doubt the toxic effects are that generalized.

Dichlorodiphenyltrichloroethane

Dichlorodiphenyltrichloroethane or DDT DDT is an organochlorine insecticide used mainly to control mosquito-borne malaria; use on crops has generally been replaced by less persistent insecticides. It was extensively used during the Second World War among Allied troops and certain civilian populations to control insect typhus and malaria vectors, and was then extensively used as an agricultural insecticide after 1945. DDT was banned for use in Sweden in 1970 and in the United States in 1972.

It is available in several different forms: aerosols, dustable powders, emulsifiable concentrates, granules and wettable powders. It is reported to be compatible with many other pesticides and incompatible with alkaline substances. Many insect pests may have developed resistance to DDT.

Unless otherwise specified, the toxicological, environmental effects and environmental fate and chemistry data refer to the technical product DDT. Technical grade DDT is actually a mixture of three isomers of DDT, principally the p,p'-DDT isomer (ca. 85%), with the o,p'-DDT and o,o'-DDT isomers typically present in much lesser amounts.

Toxicological Effects

Acute Toxicity

DDT is moderately to slightly toxic to studied mammalian species via the oral route. Reported oral LD50s range from 113 to 800 mg/kg in rats (1-3); 150-300 mg/kg in mice; 300 mg/kg in guinea pigs; 400 mg/kg in rabbits; 500-750 mg/kg in dogs and greater than 1,000 mg/kg in sheep and goats. Toxicity will vary according to formulation. DDT is readily absorbed through the gastrointestinal tract, with increased absorption in the presence of fats.

One-time administration of DDT to rats at doses of 50 mg/kg led to decreased thyroid function and a single dose of 150 mg/kg led to increased blood levels of liver-produced enzymes and changes in the cellular chemistry in the central nervous system of monkeys. Single doses of 50-160 mg/kg produced tremors in rats, and single doses of 160 mg/kg produced hind leg paralysis in guinea pigs. Mice suffered convulsions following a one-time oral dose of 200 mg/kg. Single administrations of low doses to developing 10- day old mice are reported to have caused subtle effects on their neurological development.

DDT is slightly to practically non-toxic to test animals via the dermal route, with reported dermal LD50s of 2,500-3,000 mg/kg in female rats (1-3), 1000 in guinea pigs and 300 in rabbits. It is not readily absorbed through the skin unless it is in solution.

It is thought that inhalation exposure to DDT will not result in significant absorption through the lung alveoli (tiny gas-exchange sacs) but rather that it is probably trapped in mucous secretions and swallowed by exposed individuals following the tracheo-bronchial clearance of secretions by the cilia.

Acute effects likely in humans due to low to moderate exposure may include nausea, diarrhea, increased liver enzyme activity, irritation (of the eyes, nose or throat), disturbed gait, malaise and excitability; at higher doses, tremors and convulsions are possible. While adults appear to tolerate moderate to high ingested doses of up to 280 mg/kg, a case of fatal poisoning was seen in a child who ingested one ounce of a 5% DDT:kerosene solution.

Chronic Toxicity

DDT has caused chronic effects on the nervous system, liver, kidneys,and immune systems in experimental animals. Effects on the nervous system observed in test animals include: tremors in rats at doses of 16-32 mg/kg/day over 26 weeks; tremors in mice at doses of 6.5-13mg/kg/day over 80-140 weeks; changes in cellular chemistry in the central nervous system of monkeys at doses of 10 mg/kg/day over 100 days, and loss of equilibrium in monkeys at doses of 50 mg/kg/day for up to 6 months.

The main effect on the liver seen in animal studies was localized liver damage. This effect was seen in rats given 3.75 mg/kg/day over 36 weeks, rats exposed to 5 mg/kg/day over 2 years and dogs at doses of 80 mg/kg/day over the course of 39 months. In many cases lower doses produced subtle changes in liver cell physiology, and in some cases higher doses produced more severe effects. In mice doses of 8.33 mg/kg/day over 28 days caused increased liver weight and increased liver enzyme activity. Liver enzymes are commonly involved in detoxification of foreign compounds, so it is unclear whether increased liver enzyme activity in itself would constitute an adverse effect. In some species (monkeys and hamsters), doses as high as 8-20 mg/kg/day caused no observed adverse effects over exposure periods as long as 3.5-7 years.

Kidney effects observed in animal studies include adrenal gland hemorrhage in dogs at doses of 138.5 mg/kg/day over 10 days and adrenal gland damage at 50 mg/kg day over 150 days in dogs. Kidney damage was also seen in rats at doses of 10 mg/kg/day over 27 months.

Immunological effects observed in test animals include: reduced antibody formation in mice following administration of 13 mg/kg/day for 3-12 weeks and reduced levels of immune cells in rats at doses of 1 mg/kg/day. No immune system effects were observed in mice at doses of 6.5 mg/kg/day for 3-12 weeks.

Dose levels at which effects were observed in test animals are very much higher than those which may be typically encountered by humans. The most significant source of exposure to individuals in the United States is occupational, occurring only to those who work or worked in the production or formulation of DDT products for export. Analysis of U. S. market basket surveys showed approximately a 30-fold decrease in detected levels of DDT and metabolites in foodstuffs from 1969-1974, and another three-fold drop from 1975-1981, with a final estimated daily dose of approximately 0.002 mg/person/day. Based on a standard 70-kg person, this results in a daily intake of approximately 0.00003 mg/kg/day. Due to the persistence of DDT and its metabolites in the environment, very low levels may continue to be detected in foodstuffs grown in some areas of prior use. It has been suggested that, depending on patterns of international DDT use and trade, it is possible that dietary exposure levels may actually increase over time. Persons eating fish contaminated with DDT or metabolites may also be exposed via bioaccumulation of the compound in fish.

Even though current dietary levels are quite low, past and current exposures may result in measurable body burdens due to its persistence in the body.

Adverse effects on the liver, kidney and immune system due to DDT exposure have not been demonstrated in humans in any of the studies which have been conducted to date.

Reproductive Effects

There is evidence that DDT causes reproductive effects in test animals. No reproductive effects were observed in rats at doses of 38 mg/kg/day administered at days 15-19 of gestation. In another study in rats, oral doses of 7.5 mg/kg/day for 36 weeks resulted in sterility. In rabbits, doses of 1 mg/kg/day administered on gestation days 4-7 resulted in decreased fetal weights and 10 mg/kg/day on days 7-9 of gestation resulted in increased resorptions. In mice, doses of 1.67 mg/kg/day resulted in decreased embryo implantation and irregularities in the estrus cycle over 28 weeks. It is thought that many of these observed effects may be the result of disruptions in the endocrine (hormonal) system.

Available epidemiological evidence from two studies does not indicate that reproductive effects have occurred in humans as a result of DDT exposure. No associations between maternal blood levels of DDT and miscarriage nor premature rupture of fetal membranes were observed in two separate studies. One study did report a significant association between maternal DDT blood levels and miscarriage, but the presence of other organochlorine chemicals (e.g., PCBs) in maternal blood which may have accounted for the effect make it impossible to attribute the effect to DDT and its metabolites.

Teratogenic Effects

There is evidence that DDT causes teratogenic effects in test animals as well. In mice, maternal doses of 26 mg/kg/day DDT from gestation through lactation resulted in impaired learning performance in maze tests. In a two-generational study of rats, 10 mg/kg/day resulted in abnormal tail development. Epidemiological evidence regarding the occurance of teratogenic effects as a result of DDT exposure are unavailable. It seems unlikely that teratogenic effects will occur in humans due to DDT at likely exposure levels.

Mutagenic Effects

The evidence for mutagenicity and genotoxicity is contradictory. In only 1 out of 11 mutagenicity assays in various cell cultures and organisms did DDT show positive results. Results of in vitro and in vivo genotoxocity assays for chromosomal aberrations indicated that DDT was genotoxic in 8 out of 12 cases, and weakly genotoxic in 1 case.

In humans, blood cell cultures of men occupationally exposed to DDT showed an increase in chromosomal damage. In a separate study, significant increases in chromosomal damage were reported in workers who had direct and indirect occupational exposure to DDT. Thus it appears that DDT may have the potential to cause genotoxic effects in humans, but does not appear to be strongly mutagenic. It is unclear whether these effects may occur at exposure levels likely to be encountered by most people.

Carcinogenic Effects

The evidence regarding the carcinogenicity of DDT is equivocal. It has been shown to cause increased tumor production (mainly in the liver and lung) in test animals such as rats, mice and hamsters in some studies but not in others. In rats, liver tumors were induced in three separate studies at doses of 12.5 mg/kg/day over periods of 78 weeks to life, and thyroid tumors were induced at doses of 85 mg/kg/day over 78 weeks. In mice, lifetime doses of 0.4 mg/kg/day resulted in lung tumors in the second generation and leukemia in the third generation; liver tumors were induced at oral doses of 0.26 mg/kg/day in two separate studies over several generations. In hamsters, significant increases in adrenal gland tumors were seen at doses of 83 mg/kg/day in females (but not males) , and in males (but not females) at doses of 40 mg/kg/day.

In other studies, however, no carcinogenic activity was observed in rats at doses less than 25 mg/kg/day; no carcinogenic activity was seen in mice with at doses of 3-23 mg/kg/day over an unspecified period, and in other hamster studies there have been no indications of carcinogenic effects.

The available epidemiological evidence regarding DDT's carcinogenicity in humans, when taken as a whole, does not suggest that DDT and its metabolites are carcinogenic in humans at likely dose levels. In several epimiological studies, no significant associations were seen between DDT exposure and disease, but in one other study, a weak association was observed. In this latter study, which found a significant association between long-term, high DDT exposures and pancreatic cancers in chemical workers, there were questions raised as to the reliability of the medical records of a large proportion of the cancer cases.

Organ Toxicity

Acute human exposure data and animal studies reveal that DDT can affect the nervous system, liver, kidney. Increased tumor production in the liver and lung has been observed in test animals. An association with pancreatic cancer was suggested in humans in one study.

Fate in Humans and Animals

DDT is very slowly transformed in animal systems. Initial degradates in mammalian systems are 1,1-dichloro-2,2-bis(p-dichlorodiphenyl)ethylene (DDE) and 1,1-dichloro-2,2-bis(p-chlorophenyl)ethane (DDD), which are very readily stored in fatty tissue. These compounds in turn are ultimately transformed into bis(dichlorodiphenyl) acetic acid (DDA) via other metabolites at a very slow rate. DDA, or conjugates of DDA, are readily excreted via the urine.

Available data from analysis of human blood and fat tissue samples collected in the early 1970s showed detectable levels in all samples, but a downward trend in the levels over time. Later study of blood samples collected in the latter half of the 1970s showed that blood levels were declining further, but DDT or metabolites were still seen in a very high proportion of the samples. Levels of DDT or metabolites may occur in fatty tissues (e.g. fat cells, the brain, etc.) at levels of up to several hundred times that seen in the blood. DDT or metabolites may also be elminated via mother's milk by lactating women.

Ecological Effects

Effects on Birds

DDT may be slightly toxic to practically non-toxic to birds. Reported dietary LD50s range from greater than 2,240 mg/kg in mallard, 841 mg/kg in Japanese quail and 1,334 mg/kg in pheasant. Other reported dietary LD50s in such species as bobwhite quail, California quail, red-winged blackbird, cardinal, house sparrow, blue jay, sandhill crane and clapper rail also indicate slight toxicity both in acute 5-day trials and over longer periods of up to 100 days. In birds, exposure to DDT occurs mainly through the food web through predation on aquatic and terrestrial species having body burdens of DDT, such as fish, earthworms and other birds.

There has been much concern over chronic exposure of bird species to DDT and effects on reproduction, especially eggshell thinning and embryo deaths. The mechanisms of eggshell thinning are not fully understood. It is thought that this may occur from the major metabolite, DDE, and that predator species of birds are the most sensitive to these effects. Laboratory studies on bird reproduction have demonstrated the potential of DDT and DDE to cause subtle effects on courtship behavior, delays in pairing and egg laying and decreases in egg weight in ring doves and Bengalese finches. The implications of these for long-term survival and reproduction of wild bird species is unclear.

There is evidence that synergism may be possible between DDT's metabolites and organophosphate (cholinesterase-inhibiting) pesticides to produce greater toxicity to the nervous system and higher mortality. Aroclor (polychlorinated biphenyls, or PCBs) may result in additive effects on eggshell thinning.

Effects on Aquatic Species

DDT is very highly toxic to many aquatic invertebrate species. Reported 96-hour LC50s in various aquatic invertebrates (e.g., stoneflies, midges, crayfish, sow bugs) range from 0.18 ug/L to 7.0 ug/L, and 48-hour LC50s are 4.7 ug/L for daphnids and 15 ug/L for sea shrimp. Other reported 96-hour LC50s for various aquatic invertebrate species are from 1.8 ug/L to 54 ug/L. Early developmental stages are more susceptible than adults to DDT's effects. The reversibility of some effects, as well as the development of some resistance, may be possible in some aquatic invertebrates.

DDT is very highly toxic to fish species as well. Reported 96-hour LC50s are less than 10 ug/L in coho salmon (4.0 ug/L), rainbow trout (8.7 ug/L), northern pike (2.7 ug/L), black bullhead (4.8 ug/L), bluegill sunfish (8.6 ug/L), largemouth bass (1.5 ug/L), and walleye (2.9 ug/L). The reported 96-hour LC50s in fathead minnow and channel catfish are 21.5 ug/L and 12.2 ug/L respectively. Other reported 96-hour LC50s in largemouth bass and guppy were 1.5 ug/L and 56 ug/L respectively. Observed toxicity in coho and chinook salmon was greater in smaller fish than in larger. It is reported that DDT levels of 1 ng/L in Lake Michigan were sufficient to affect the hatching of coho salmon eggs. DDT may be moderately toxic to some amphibian species and larval stages are probably more susceptible than adults.

In addition to acute toxic effects, DDT may bioaccumulate significantly in fish and other aquatic species, leading to long-term exposure. This occurs mainly through uptake from sediment and water into aquatic flora and fauna, and also fish. Fish uptake of DDT from the water will be size-dependent with smaller fish taking up relatively more than larger fish. A half- time for elimination of DDT from rainbow trout was estimated to be 160 days.

The reported bioconcentration factor for DDT is 1,000 to 1,000,000 in various aquatic species, and bioaccumulation may occur in some species at very low environmental concentrations. Bioaccumulation may also result in exposure to species which prey on fish or other aquatic organisms (e.g., birds of prey).

Effects on other Animals (Nontarget Species)

Earthworms are not susceptible to acute effects of DDT and its metabolites at levels higher than those likely to be found in the environment, but they may serve as an exposure source to species that feed on them. DDT is non-toxic to bees; the reported topical LD50 for DDT in honeybees is 27 ug/bee. Laboratory studies indicate that bats may be affected by DDT released from stored body fat during long migratory periods.

Environmental Fate

Breakdown in Soil and Groundwater

DDT is very highly persistent in the environment, with a reported half life of between 2-15 years and is immobile in most soils. Routes of loss and degradation include runoff, volatilization, photolysis and biodegradation (aerobic and anaerobic). These processes generally occur only very slowly. Breakdown products in the soil environment are DDE and DDD, which are also highly persistent and have similar chemical and physical properties.

Due to its extremely low solubility in water, DDT will be retained to a greater degree by soils and soil fractions with higher proportions of soil organic matter. It may accumulate in the top soil layer in situations where heavy applications are (or were) made annually; e.g., for apples. Generally DDT is tightly sorbed by soil organic matter, but it (along with its metabolites) has been detected in many locations in soil and groundwater where it may be available to organisms. This is probably due to its high persistence; although it is immobile or only very slightly mobile, over very long periods of time it may be able to eventually leach into groundwater, especially in soils with little soil organic matter.

Residues at the surface of the soil are much more likely to be broken down or otherwise dissipated than those below several inches. Studies in Arizona have shown that volatilization losses may be significant and rapid in soils with very low organic matter content (desert soils) and high irradiance of sunlight, with volatilization losses reported as high as 50% in 5 months. In other soils this rate may be as low as 17- 18% over 5 years. Volatilization loss will vary with the amount of DDT applied, proportion of soil organic matter, proximity to soil-air interface and the amount of sunlight.

Breakdown of Chemical in Surface Water

DDT may reach surface waters primarily by runoff, atmospheric transport, drift, or by direct application (e.g. to control mosquito-borne malaria). The reported half-life for DDT in the water environment is 56 days in lake water and approximately 28 days in river water. The main pathways for loss are volatilization, photodegradation, adsorption to water-borne particulates and sedimentation. Aquatic organisms, as noted above, also readily take up and store DDT and its metabolites. Field and laboratory studies in the United Kingdom demonstrated that very little breakdown of DDT occurred in estuary sediments over the course of 46 days.

DDT has been widely detected in ambient surface water sampling in the United States at a median level of 1 ng/L (part per trillion).

Breakdown of Chemical in Vegetation

DDT does not appear to be taken up or stored by plants to a great extent. It was not translocated into alfalfa or soybean plants, and only trace amounts of DDT or its metabolites were observed in carrots, radishes and turnips all grown in DDT-treated soils. Some accumulation was reported in grain, maize and riceplants, but little translocation occured and residues were located primarily in the roots.

References

- Polychlorinatedbiphenyls, envhealth-factsheets: idph.state.il.us, Retrieved 29 June, 2020
- Arsenic, news-room-fact-sheets-detail: who.int, Retrieved 08 May, 2020
- Rout, Gyana Ranjan; Das, Premananda (2009). "Effect of Metal Toxicity on Plant Growth and Metabolism: I. Zinc". In Lichtfouse, Eric; Navarrete, Mireille; Debaeke, Philippe; Véronique, Souchere; Alberola, Caroline (eds.). Sustainable Agriculture. pp. 873–84. doi:10.1007/978-90-481-2666-8_53. ISBN 978-90-481-2666-8
- Benzene, ipcs-features: who.int, Retrieved 09 August, 2020
- A critical review of biomarkers used for monitoring human exposure to lead: advantages, limitations, and future needs. Environ Health Perspect. 2005;113:1669–74
- Formaldehyde, sites-production-files-documents: epa.gov, Retrieved 15 April, 2020
- Mercury exposure and public health. Pediatr Clin North Am. 2007;54(2):237–269
- Bernhoft, RA (2012). "Mercury toxicity and treatment: a review of the literature". Journal of Environmental and Public Health. 2012: 460508. doi:10.1155/2012/460508. PMC 3253456. PMID 22235210

3
Ecotoxicology

Ecotoxicology is an integrated field of toxicology and ecology that studies the effects of toxic chemicals on biological organisms. A few of its concepts include microtox bioassay, dead zone, persistent organic pollutant, soil ecotoxicology, etc. This chapter discusses the subject of ecotoxicology in detail.

Ecotoxicology deals with the fate and effects of contaminants in the biosphere. Contrary to human toxicology, which is concerned with effects of chemicals on the individual organism, ecotoxicology is concerned with toxic effects on ecological entities, i.e. populations, communities or ecosystems. In practice, however, ecotoxicology builds on conventional toxicological endpoints which are determined in selected sentinel species or in species representing trophic levels in a food chain, and infers from these individual-level effects to processes at higher levels of biological organization. This approach inherently suffers from uncertainty and ignorance.

Changes of developmental and reproductive parameters can be major drivers of alterations in population growth. Therefore, toxicant effects on development and reproduction are of particular relevance when assessing the risk of environmental pollution to ecological structures and functions. The field has taken an impressive development over the last years, which has been stimulated at least in part by the emergence of the endocrine disruption issue. There is compelling evidence that hormonally active substances have adverse impacts on wild fish populations, and since endocrine-disrupting chemicals (EDCs) affect particularly developmental and reproductive properties, this has necessitated, on the one hand, a better understanding of the involved mechanisms and, on the other hand, the development of appropriate testing methodologies and assessment strategies. A further momentum to the recent progress of developmental and reproductive toxicity of fishes comes from the increasing use of small laboratory fish species as models in biomedical, toxicological and pharmacological research – a development ecotoxicology is clearly taking advantage of this approach.

In ecotoxicology, models will often be confronted with experimental or monitoring data. This can be done in a calibration process (to fit model parameters to observations) or in a validation attempt (comparing model predictions to independent observations).

First, it is important to distinguish between quantal (discrete) and graded (continuous) responses. In a quantal response, only the presence or absence of a response is determined on each individual (eg, whether it is dead or alive). As a consequence, the LC50 thus represents the concentration at which half of the test population is expected to die (after a specified exposure duration). In contrast, a graded response implies that the degree of response in each individual is determined (eg, the number of eggs produced per female). An EC50 for reproduction thus represents the predicted concentration at which the average female shows a 50% reduction in egg production. These types of responses require different models and different statistical procedures in their confrontation with data.

In model calibration, we need to judge the deviations between model and observations, such that we can prefer one fit over another and find the parameters that yield the "best possible fit" to the data. Deviations are inevitable as observations contain error and the model is always "wrong" (as it is a simplification of reality). We thus also need a model for these deviations (or residuals) and that is the area of statistics. The most-commonly applied statistical model is to assume that the deviations are random samples from independent normal distributions with mean zero and a constant variance. This set of assumptions underlies the popular least-squares analysis. However, in many practical cases, these assumptions will be violated. Quantal responses, such as survival, are discrete responses; assuming that the residuals follow a normal distribution is therefore a questionable practice (especially with small numbers of individuals). Body weight of an organism is a graded response, and a normal distribution may be more appropriate. However, a constant variance is questionable (large body weights tend to show more variation), and if we measure the same individuals over time, independence is violated. To what extent such violations affect the interpretation of a data set is not always clear. In any case, it is good to explicitly mention the assumptions made for the statistical model, just as for the dynamic model that describes the process.

When we have selected a statistical model for the residuals, we can fit the model to the data and calculate confidence intervals on the parameter estimates. It is essential to realize that a confidence interval is only meaningful when both the process model and the statistical model are meaningful (the interval is "true" given that the models are "true"). In practice, no model is ever "true," which implies that all confidence intervals should be regarded as approximations. A wide confidence interval can mean several things: the parameter is not (substantially) influencing the model output, or the data provide no (or insufficient) information on this parameter, or the parameter is heavily correlated to another one (fixing the other parameter could yield a tight interval for the first parameter). A tight confidence interval implies that none of these issues are at play. Several techniques are available to construct confidence intervals, which may yield different intervals.

Modelers will often be confronted by the question of whether their model is "validated." But what is meant by a validated model? Model validation is the process by which model outputs are (systematically) compared to independent real-world observations to judge the quantitative and qualitative correspondence with reality. A validated model is therefore a model for which this comparison has been performed, but it says nothing about the outcome. There is no strict distinction between models that are, and those that are not, "valid"; much will depend on the purpose for which the model is to be used.

Biochemical ecotoxicology is the study of the effects of contaminants in ecosystems using biochemical methods. The scope of this discipline is vast and complex because it implicates exposure to low concentrations of mixtures in the long term (chronic), encompassing most of the life stages of organisms. Although the toxicity of chemicals mostly depends on the physicochemical properties of chemicals, the resulting effects will depend on the lifestyle of the organisms, such as nutritional status, reproduction activity, prey–predation pressures, and habitat characteristics. The advent of climate change adds another level of complexity to predicting the toxic outcomes of pollutants, which is a challenge for the risk evaluation community. The biochemical approaches in ecotoxicology have the advantage of finding early warning signals at the molecular level, which can lead to effects at higher levels of biological organization. The interactions of chemicals with biochemical targets in cells or tissues that could lead to altered function represent a special type of "biomarker" in toxicology. There are many definitions proposed in the literature for biomarkers, but we think this one best reflects the biochemical markers: the term biomarker is used in a broad sense to include almost any measurement reflecting an interaction between a biological system and an environmental agent, which may be of chemical, physical, or biological origin.

In ecotoxicology, a series of biomarkers are often measured along with bioaccumulation measurements of various contaminants in tissues. This leads to a complex dataset where the relationship between each endpoint becomes unclear and also if some or all of the above measurements are characteristic to the treatment group or site location (spatial survey) or point of time (temporal survey). This situation is often observed with field-based studies or studies dealing with high-throughput assays such as the "omics" approaches.

Factorial Analysis

Factorial analysis is performed when one wants to find which biomarkers (or a subgroup) best explain the variance of all biomarkers in a complex dataset. For example, if a research group measured the expression of 50 genes in mussels collected at 10 sites upstream and downstream an urban area, one could ask whether a subgroup of gene expression changes could explain the overall responses and whether some genes are more important than others. In other words, is there a subgroup of endpoints that is correlated with n-dimensional vector composed with the 50 genes of interest? This analysis could be considered as a data reduction method in finding which biomarkers best explain the whole dataset. In some cases, no subgroups of biomarkers could be found and the whole dataset should be kept to examine the changes across 10 sites. In many instances, a subset of biomarkers could be proposed that would reduce cost and time to study the effects across the 10 sites in an urban area. Factorial analysis is also a methodology to examine the inter-relationships between biomarkers, which could provide interesting insights to better understand the interaction between the toxicity of pollution at the biochemical/physiological level.

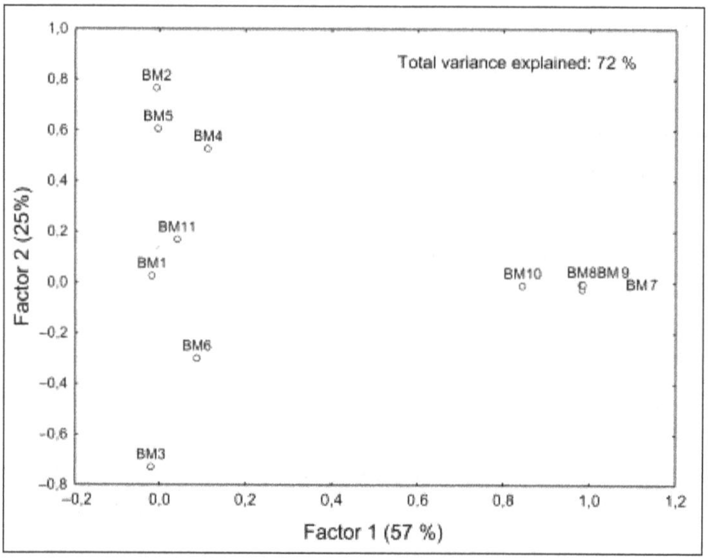

Example of factorial analysis results using the principal component procedure.

This example is based on the production of 11 biomarkers in organisms exposed to various treatments or collected at different sites. The biomarkers could be contaminant tissue burdens and any biomarkers to determine organism's health and toxicity. Factorial analysis using the principal component procedure is the most commonly used method, but other methods can be used, such as centroid, principal axis, and multiple R^2. These methods for extraction of main factors are available in many types of statistical software. The method determines principal factors, which are eigenvalues obtained from *n*endpoints (biomarkers). The first two factors, which explain most of variance, are plotted on an *X* and *Y* plot. The percentage value on the *X*- and *Y*-axis are the proportion of the variance explained where the total is indicated in the plot (72%). In practice, if the variance is explained at more than 60% then the analysis could be considered acceptable for identifying the most relevant biomarkers. If the total explained variance is below 60% then the biomarker used cannot explain all the variations in the data. In the figure, factorial weights (or the correlation of the biomarker with factors 1 and 2) of the biomarkers (BM1–11) are plotted in a 2D plot. Biomarkers with factorial weights >70% are considered the most important biomarkers or principal components because they are more strongly correlated with the eigenvalue of factors 1 and 2. In the example, these are BM2, 3, and 7–10. This indicates that 72% of the data could be explained by 6 of the initial 11 biomarkers. In these plots, biomarkers grouped together (e.g., BM 7–10) are generally positively correlated with each other while the ones that are diametrically opposed (e.g., BM2 and 3) are often negatively correlated. This provides information on the inter-relationships between biomarkers; for example, if we examined the biomarker of toxicity in mussels collected at 10 sites upstream and downstream a polluted area—the biomarkers that are related to polyaromatic hydrocarbons in tissues and biotransformation activity show high factorial weights—then the analysis suggests that these compounds are the main culprit in the contamination source. Of course, this is defendable if other types of biomarkers were examined.

Discriminant Function Analysis

Discriminant function analysis is similar to multivariate ANOVA but indicates how well the treatment groups or study sites differ with each other. Discriminant analysis seeks out a linear combination of biomarker data for each treatment group that maximizes the difference between treatment groups or study sites for proper classification. The analysis provides a classification function that determines to which groups an individual belongs:

$$s_i = c_i + wi1 * x_1 + w_i2 * x_2 + ... + w_{im} * x_m$$

where i represents the respective groups or sites and numbers 1,2, ..., m represent the m biomarkers (variables). c_i is the constant for group i and w_{ij} is the ponderation factor of variable (biomarkers) j for group i. S_i is the classification value. Different sites will give classification efficiency greater than 70% while similar sites would give undistinguishable results (<70% classification). The analysis provides a classification matrix that shows how well each individual or treatment group replicate is associated to a given site or treatment group. The results for discriminant function analysis are usually expressed by the Wilks' lambda statistic and displayed by a 2D plot of the two best discriminant functions in respect to site classification. The analysis also provides factorial analysis to identify which biomarkers are strongly correlated with the discriminant functions 1 and 2. In the example, 11 biomarkers were examined in six sites in local mussel populations. In this analysis, each data point represents the root function of the combined 11 biomarkers for each individual. The data points are also displayed for each treatment group or site of collection. In the legend, the classification results are presented in percentage correctness. For example, group or site 2 was correctly classified at 100% but sites 3 and 5 were misclassified as site 2. This indicates that sites 2, 3, and 5 could be considered a similar "composite site," or these sites could be considered similar in their ecotoxicological properties. Group 6 was correctly classified at 100% although close to site/group 1. The biomarkers that had the highest correlation with the root function are listed in parentheses on the x- and y-axes. This type of analysis is often used in ecotoxicology because it provides a great deal of information for data mining and analysis. Indeed, it provides information on how sites differ with each other and which biomarkers contribute to site classification/discrimination.

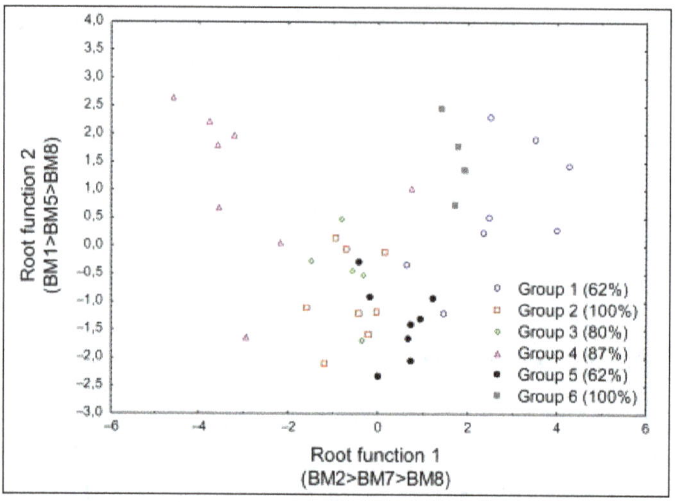

Example of discriminant function analysis for site classification. Eleven biomarkers (BM) were determined in six groups (sites or treatments) and analyzed by discriminant function analysis. Each data point corresponds to each replicate individual in a group. The percentage values of groups 1–6 represent the classification correctness. Groups that overlap with each other are considered similar, while distinct groups are considered different. BMs that were the most strongly correlated with the root functions 1 and 2 are mentioned on the *x*- and *y*-axis.

In discriminant function analysis, the factorial scores could be calculated for each observation, which is a linear combination of the standardized biomarker measured for each individual. It provides an integrated metric that reflects the changes in the 11 biomarkers tested. The larger the factorial weights (absolute value) the greater the contribution of the biomarker to the root functions 1 and 2.

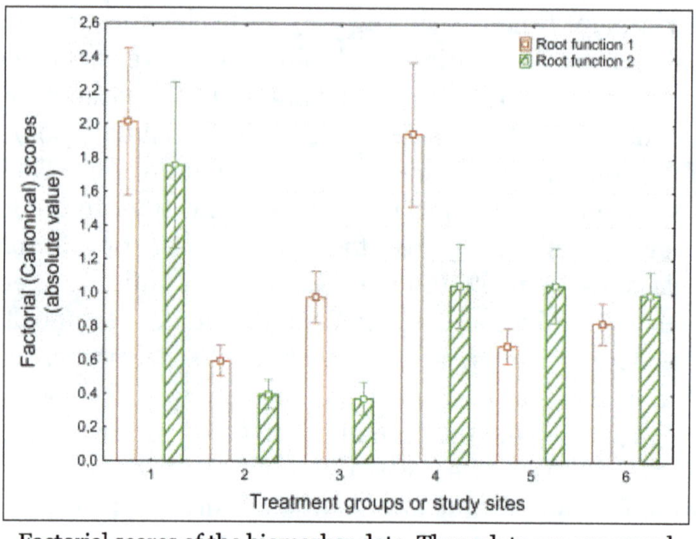

Factorial scores of the biomarker data. These data are expressed as the mean of the absolute scores with the SE.

Ecological Effects of Toxic Substances

Chemicals released into the environment may have a variety of adverse ecological effects. Ranging from fish and wildlife kills to forest decline, ecological effects can be long-term or short-lived changes in the normal functioning of an ecosystem, resulting in economic, social, and aesthetic losses. These potential effects are an important reason for regulation of pesticides, toxic substances, and other sources of pollution.

The physical environment along with the organisms (biota) inhabiting that space, make up an ecosystem. Some typical examples of ecosystems include: a farm pond, a mountain meadow, and a rain forest.

An ecosystem follows a certain sequence of processes and events through the days, seasons, and years. The processes include not only the birth, growth, reproduction, and death of biota in that particular ecosystem, but also the interactions between species and physical characteristics of the geological environment. From these processes the ecosystem gains a recognizable structure and function, and matter and energy are cycled and flow through the system. Over time, better adapted species come to dominate; entirely new species may change, perhaps in a new or altered ecosystem.

Organization of Ecosystems

The basic level of ecological organization is with the individual, a single plant, insect or bird. The definition of ecology is based on the interactions of organisms with their environment. In the case of an individual, it would entail the relationships between that individual and numerous physical (rain, sun, wind, temperature, nutrients, etc.) and biological (other plants, insects, diseases, animals, etc.) factors. The next level of organization is the population. Populations are no more than a collection of individuals of the same species within an area or region. We can see populations of humans, birch trees, or sunfish in a pond. Population ecology is concerned with the interaction of the individuals with each other and with their environment.

The next, more complex, level of organization is the community. Communities are made up of different populations of interacting plants, animals, and microorganisms also within some defined geographic area. Different populations within a community interact more among themselves than with populations of the same species in other communities, therefore, there are often genetic differences between members of two different communities. The populations in a community have evolved together, so that members of that community provide resources (nutrition, shelter) for each other.

The next level of organization is the ecosystem. An ecosystem consist of different communities of organisms associated within a physically defined space. For example, a forest ecosystem consists of animal and plant communities in the soil, forest floor, and forest canopy, along the stream bank and bottom, and in the stream. A stream bottom community, for example, will have various fungi and bacteria living on dead leaves and animal wastes, protozoans and microscopic invertebrates feeding on these microbes, and larger invertebrates (worms, crayfish) and vertebrates (turtles, catfish). Each community functions somewhat separately, but are also linked to the others by the forest, rainfall, and other interactions. For example, the stream community is heavily dependent upon leaves produced in the surrounding trees falling into the stream, feeding the microbes and other invertebrates. For another example, the rainfall and groundwater flow in a surrounding forest community greatly affects the amount and quality of water entering the stream or lake system.

Terrestrial ecosystems can be grouped into units of similar nature, termed biomes (such as a "deciduous forest," "grassland," "coniferous forest," etc.), or into a geographic unit, termed landscapes, containing several different types of ecosystems. Aquatic ecosystems are commonly categorized on the basis of whether the water is moving (streams, river basins) or still (ponds, lakes, large lakes) and whether the water is fresh, salty (oceans), or brackish (estuaries). Landscapes and biomes (and large lakes, river basins, and oceans) are subject to global threats of pollution (acid deposition, stratospheric ozone depletion, air pollution, the greenhouse effect) and human activities (soil erosion, deforestation).

Adverse Effects on Ecosystems

While many natural forces -- drought, fire, flood, frost or species migration -- can affect it, an ecosystem will usually continue to function in a recognizable way. For instance, a pond ecosystem may go through flood or drought but continues to be a pond. This natural resilience of ecosystems enables them to resist change and recover quickly from disruption.

On the other hand, toxic pollutants and other non-natural phenomena can overwhelm the natural stability of an ecosystem and result in irreversible changes and serious losses, as illustrated by the following examples:

- Decline of forests, due to air pollution and acid deposition.
- Loss of fish production in a stream, due to death of invertebrates from copper pollution.
- Loss of timber growth, due to nutrient losses caused by mercury poisoning of microbes and soil insects.
- Decline and shift in age of eagle and hawk (and other top predator) populations, due to the effects of ddt in their food supply on egg survival.
- Losses of numbers of species (diversity) in ship channels subjected to repeated oil spills.
- Loss of commercially valuable salmon and endangered species (bald eagle, osprey) from forest applications of DDT.

Each of these pollutant-caused losses has altered ecosystem processes and components and thus affected aesthetic and commercial value of an ecosystem.

Usually, adverse ecological effects take place over long period of time or even at some distance from the point of release of a chemical. For example, DDT, though banned for use in the United States for over twenty years, is still entering the Great Lakes ecosystem through rainfall and dust from sources half way around the world. The long-term effects and overall impacts of new and existing chemicals on ecosystems can only be partially evaluated by current laboratory testing procedures. Nevertheless, through field studies and careful monitoring of chemical use and biological outcome, it is possible to evaluate the short-term and long-term effects of pesticides and other chemicals.

Adverse Ecological Effects on Communities

Scientists are most concerned about the effects of chemicals and other pollutants on communities. Short-term and temporary effects are much more easily measured than long-term effects of pollutants on ecosystem communities. Understanding the impact of effects requires knowledge of the time course and variability of these short-term changes.

Pollutants may adversely affect communities by disrupting their normal structure and delicate interdependencies. The structure of a community includes its physical system, usually created by the plant life and geological processes, as well as the relationships between its populations of biota.

For example, a pollutant may eliminate a species essential to the functioning of the entire community; it may promote the dominance of undesirable species (weeds, trash fish); or it may simply decrease the numbers and variety of species present in the community. It may also disrupt the dynamics of the food webs in the community by breaking existing dietary linkages between species. Most of these adverse effects in communities can be measured through changes in productivity in the ecosystem. Under natural stresses (for example, unusual temperature and moisture conditions), the community may be unable to tolerate effects of a chemical otherwise causing no harm.

An important facet of biological communities is the number and intensity of interactions between species. These interactions make the community greater than simply the sum of its parts. The community is stronger than its populations, and the ecosystem is more stable than its communities. A seriously altered interaction may adversely affect all the species dependent on it. Even so, some ecosystem properties or functions (such as nutrient dynamics) can be altered by chemicals without apparent effects on populations or communities. Thus, an important part of research in ecological effects is concerned with the relative sensitivity of ecosystems, communities, and populations to chemicals and to physical stresses.

Consider the effects of spraying an orchard with an insecticide when bees and other beneficial insects may be present and vulnerable to the toxicant. This practice is both economically and ecologically unsound, since it would deprive all plants in the area of pollinators and disrupt control of plant pests by their natural enemies. Advanced agricultural practices, such as integrated pest management (IPM), avoid these adverse effects through appropriate timing and selection of sprays in conjunction with non-chemical approaches to insect control.

Effects of chemicals on communities can be measured in laboratory model ecosystem (microcosm) studies, in intermediate sized systems (mesocosms, engineered field systems, open-top plant chambers, field pens), and in full field trials. Thus, data gathered about effects of chemicals on processes and species can be evaluated in various complex situations that reflect the real world.

Adverse Effects on Species

Most information on ecological effects has been obtained from studies on single species of biota. These tests have been performed in laboratories under controlled conditions and chemical exposures, usually with organisms reared in the laboratory representing inhabitants of natural systems. Most tests are short-term, single exposures (acute toxicity assays), but long-term (chronic) exposures are used as well. Although such tests reveal which chemicals are relatively more toxic, and which species are relatively more vulnerable to their effects, these tests do not disclose much about either the important interactions noted above or the role of the range of natural conditions faced by organisms in the environment.

Generally, the effects observed in these toxicity tests include reduced rates of survival or increased death rates; reduced growth and altered development; reduced reproductive capabilities, including birth defects; changes in body systems, including behavior; and genetic changes. Any of these effects can influence the ability of species to adapt and respond to other environmental stresses and community interactions.

Environmental toxicology studies performed on species in the laboratory provide the basis for much of the current regulation of pollutants and have allowed major improvements in environmental quality. However, these tests yield only a few clues to effects on more complex systems. Long-term studies and monitoring of ecological effects of new and existing chemicals released into the environment are needed in order to create understanding of potential adverse ecological effects and their consequences.

Adverse ecological effects from environmental pollutants occur at all levels of biological organization, but most information about these effects has been obtained with single species. The effects can be global or local, temporary or permanent, or short-lived (acute) or long-term (chronic). The most serious effects involve loss in production, changes in growth, development and behavior, altered diversity or community structure, changes in system processes (such as nutrient cycling), and losses of valuable species. These ecological losses in turn may be economically, aesthetically, or socially important. Hence, ecological effects are of serious concern in regulating pollutants and a variety of tests have been devised to help evaluate the potential for adverse ecological effects. Developing an understanding of how these tests and other information can be used to prevent environmental problems caused by pollutants is the basis for ecological risk assessment research.

Ecotoxicity

Pesticide distribution.

Ecotoxicity, the subject of study of the field of ecotoxicology (a portmanteau of ecology and toxicology), refers to the potential for biological, chemical or physical stressors to affect ecosystems. Such stressors might occur in the natural environment at densities, concentrations or levels high enough to disrupt the natural biochemistry, physiology, behavior and interactions of the living organisms that comprise the ecosystem.

Oil spill creek.

Ecotoxicology has been defined as, "the branch of toxicology concerned with the study of toxic effects, caused by natural or synthetic pollutants, to the constituents of ecosystems, animal (including human), vegetable and microbial, in an integral context".

Oil spill.

Common Environmental Toxicants

- Diethyl phthalate, that enter environments through industries manufacturing cosmetics, plastic and many other commercial products.
- Bisphenol A (BPA), found in many mass-produced products such as, medical devices, food packaging, cosmetics, children's toys, computers, CD's, etc.
- Pharmaceuticals such as Climbazole, a fungicide found in anti-dandruff shampoos.
- Pesticides.
- Some but not all; cleaning products, laundry detergents, fabric softeners, oven cleaners, disinfectants.
- Phosphates.
- Oil.

Everyday Hazards

In Canada, there is no law requiring manufactures to state the health and environmental hazards associated with their cleaning products. Many people buy such products to support a clean and healthy home, often unaware of the products ability to harm both their own health and the surrounding environment. "Canadians spend more than $275 million on household cleaning products in a year" Chemicals from these cleaners enter our bodies through air passageways and absorption through the skin and when these cleaning products are washed down the drain they negatively affect aquatic ecosystems. There are also no regulations in place stating that the ingredients be listed on labels of cleaning products leading the users to be ultimately unaware of the chemicals they expose themselves and their surrounding environments to.

Fragrance Chemicals

Fragrance chemicals are found in most cleaning products, perfumes, and personal care products. More than 3000 chemicals are used in these fragrance mixtures. The synthetic musks used in detergents accumulate in the environment and are harmful to aquatic organisms. Certain musks are possible endocrine disruptors that interfere with hormone functioning. Phthalates are a common ingredient in these fragrance mixtures found in laundry detergents and fabric softeners. These phthalates are suspected endocrine disrupters that affect reproduction rates including reduced sperm count in males. Certain glass cleaners and floor polishes contain dibutyl phthalate (DBP). The European Union classifies DBP as very toxic to aquatic organisms, posing a huge danger as these cleaners, especially the floor polishes are often rinsed down the drain and into aquatic environments.

Phosphates

Phosphates are found in many dishwasher detergents, laundry detergents, and bathroom cleaners. They act as a fertilizer in water and in high concentrations can promote algae blooms and increase weed growth. When water containing phosphates are washed into water areas they carry with them fertilizers, nutrients, and wastes from the land. Phytoplankton and algae flourish at the surface due to increased phosphates. Dead phytoplankton and other organisms sink to the bottom giving rise to large numbers of decomposers due to increased food supply (dead organisms, phytoplankton). Due to the increased number of decomposers that use more oxygen, fish and shrimp at the lower layers of the ocean become oxygen-starved and hypoxic zones become apparent.

Quaternary Ammonium Compounds (Quats)

Quats are anti-microbial agents that are found in bathroom cleaners, fabric softeners, and degreasers. They are a class of irritants and sensitizers that negatively affect people who suffer from asthma. Chemicals in this class are persistent in aquatic environments, and toxic to the organisms that live in these environments. Many researchers are concerned that their widespread use in everyday household disinfectants and cosmetics are contributing to antibiotic resistant bacteria, thus limiting microbial infection treatment options.

Trisodium Nitrilotriacetate

Trisodium nitrilotriacetate is found in bathroom cleaners and possibly some laundry detergents although more actively used in industrial formulations. Small amounts add up in the environment and add to an overall toxic issue. In aquatic ecosystems these chemicals cause heavy metals in sediment to redissolve and many of these metals are toxic to fish and other wildlife.

Plasticizers

Phthalates and BPA date back to the 1920s and 1930s. Phthalates have been applied as polyvinyl chloride (PVC) additives since 1926, but were also used for health care purposes as insect repellents and cercaricides. BPA is present in most aquatic environments, entering water systems through landfills and sewage treatment plant runoff allowing for bioaccumulation in aquatic organisms. These endocrine disrupters are a large group of chemicals that enter into the aquatic environment through the manufacturing of various industrial and consumer products, agriculture and food/drug processing, waste water treatment plants and human wastes. Phthalate esters are common additives that soften and make PVC more flexible. It is used in many everyday items such as medical devices, packaging for fragrances and cosmetics, ropes and varnishes, in plastic used to wrap food, and shower curtains. These Phthalate esters have been found in areas of water, air, sediment, and in gulfs and rivers around the world, Giam as cited by. Phalates and BPA affect reproduction in animal groups such as Molluscs, crustaceans, amphibians and fish. Most of these plasticizers affect hormone systems, and some phthalates have even larger pathways of disruption. Phthalates and BPA have been proven to affect development and reproduction in a variety of species. Disturbances include changes in the number of offspring produced and reduced hatching success. In amphibians for example, phthalates and BPA disrupt thyroid functioning which in turn impacts larval development. Molluscs, crustaceans and amphibians appear to be more responsive than fish, with most effects being induced in low concentration ranges with the exception of disrupted spermatogenesis in fish in the low range.

A Phthalate referred to as diethyl phthalate (DEP) enters the aquatic environment through industries that manufacture cosmetics, plastics and many commercial products that pose hazards to aquatic organisms and human health. Through exposing an adult male common Carp to LC50 doses it was evident that a bioaccumulation of DEP in testis, liver, brain, gills and muscle tissue was present. Fish exposed to 20 ppm of DEP became drowsy and discolored during the onset of the fourth week. Sources of DEP contamination and accumulation in humans include cosmetic products and dietary meat of fish, Persky This DEP acts as a cosmetic ingredient and vehicle for fragrances, both which come in contact with the skin. Many countries around the world including India practice sewage fed fisheries where waste waters are used for the purpose of culturing fish. Endocrine disruption and a presence of phthalate residue is highly likely to be observable in these sewage fed fish. This is the case as waste water from various industries and garbage containing DEP are released into these waters. Through a DEP treatment with Cyprinus carpio, liver size was observed to increase and testis size decreased. In fish, muscle ALT and AST activities decreased as it was effected by DEP treatment. Like many toxic chemicals DEP has been known to affect metabolic enzyme profiles and activities of phosphatases and transaminases, Ghorpade as cited by. A decrease in immunity of M. rosenbergii after exposure to DEP was also noted. Given that biological effect concentrations for plasticizers used in the laboratory coincide with concentrations present in the environment, it seems that some wildlife species must be negatively impacted.

Personal Hygiene Products

Personal care products can reach the environment through drainage from waste water treatment plants and digested sludge. Recently, the anti-dandruff and antimycotic, Climbazole, was detected in waste water treatment drainages. Climbazole is readily used in cosmetics, and is an ingredient in anti-dandruff shampoos. Shampoos contain formulations of up to 2.0% which is the equivalent of approximately 15g/L. Climbazole is classified as very toxic to aquatic organisms. It affects the growth of green algae Pseudokirchneriella subcapitata at very low concentrations. Zebrafish experienced lethal effects after exposure to climbazole in laboratory testing. Effects included thickening of fertilized eggs, lack of somite formation, lack of detachment of the tail bud from the yolk sac, and lack of a heartbeat were evaluated after 48 hours. Along with Danio rerio, Lemna minor, Navicula pelliculosa, Pseudokirchneriella subcapitata, and Daphnia magna were tested and all were found to be negatively affected by climbazole in a concentration-dependent manner with highest toxicity observed in L. minor. Effects included stunted colony growth and darkening in color. Effects of climbazole on oats and turnip included retarded, stunted growth of the leaves and shoot as well as turning darker in color. The aquatic ecotoxicity of climbazole can be classified as very toxic to Lemna and algae, toxic to fish and harmful to Daphnia.

Pesticides

Pesticides often pose serious problems as they kill not only targeted organisms but also non-targeted organisms in the process. They are released into the natural environment intentionally by people who are often unaware that the chemicals will travel further than anticipated, Hatakeyama as cited in. Thus the pesticides largely affect the natural communities in which they are used. They negatively effect multiple levels ranging from molecules to tissues to organs to individuals to populations and onto communities. In the natural environment, a combination of pesticide exposure and natural stressors such as fluctuating temperature, food shortage or decreased oxygen availability are worse than when presented alone. Pesticides can affect the feeding rates of zoo-plankton. In the presence of pesticides zoo-plankton display lower feeding rates which result in reduced growth and reproduction. Swimming may also be affected by pesticides which poses a life-threatening issue for zoo-plankton as they swim to obtain food and nutrients and avoid predators. Such changes may alter predator-prey relationships. A spinning behavior became apparent in Daphnia induced by carbaryl which increased the probability of the Daphnia being eaten by other fish, Dodson as cited by. The toxicant pentachlorophenol increases swimming speed in the rotifer Brachionus calyciflorus, in turn increasing the encounter rate of the prey with their predators, Preston as cited by.

Oil Spills

One of the major environmental impacts of oil exploration on the environment is the contamination of aquatic ecosystems from oil spills and oil seepages from pits. Oftentimes, as is the case in the Ecuadorian Amazon, oil is used to control dust on roadways, causing the precipitation runoff from these roads to also be contaminated. Direct human health hazards occur since many people, including children walk barefoot on these oiled roads putting them in direct contact with the crude oil. Other hazards to humans include seepages into ponds that provide drinking water for the population. During the exploration for oil, mud that has been drilled is deposited into pits. These production pits are often not lined risking the possibility for contaminants to leak into the surrounding environment. Environmental concerns are primarily focused on a group of polycyclic aromatic hydrocarbons (PAHs). "PAHs accumulate on particles and sediments, which tend to protect them from biodegrading processes", Green and Trett as cited in. During the exploration for oil, mud that has been drilled is deposited into pits. These production pits are often not lined risking the possibility for contaminants to leak into the surrounding environment. Samples were collected from four sites (13 stations) in the Ecuadorian Amazon where crude oil was the main pollutant. The water collected from Site B, a drinking water pond located 100m from an in use pit, had the highest total petroleum hydrocarbon (THP) concentration. Sediments were found to be acutely phototoxic. This area which has poorly developed infrastructure is one where residents collect water for drinking, cooking and bathing from the rivers and ponds nearby.

"A recent study observed excess cancer rates in a village in this region" Sebastian et. Al, as cited in. Not only were excess cancer rates apparent but many people in this area that were consuming the water for drinking purposes became ill. In Wernersson's study, toxicity of water and sediment samples were studied on Daphnia magna (a crustacean zoo-plankton species) and Hyalella azteca (an amphipod). These samples were collected from four sites where crude oil was the main source of pollution. 1-4 day old organisms of both species were used in the tests. Immobility of D. magna was recorded after 24 hours of exposure indoors. They were then moved outdoors where they were exposed to sunlight. After 1–2 hours the samples were removed from the sunlight and it was found that D. Magna often recovered within an hour after UV exposure. Hyalella azteca was cultured in the same medium as was used for the D. Magna species. To minimize stress shade was provided. 16 hours of light and 8 hours of darkness were provided. Lethality was recorded after 96 hours of exposure.

Soil Ecotoxicology

Developments in soil ecotoxicology started with observations on pesticide effects on soil invertebrates in the 1960s. To support the risk assessment of chemicals, in the 1980s and 1990s development of toxicity tests was the main issue, including single species tests and also more realistic test systems like model ecosystems and field tests focusing on structural and functional endpoints. In the mean time, awareness grew about issues like bioavailability and routes of exposure, while biochemical endpoints (biomarkers) were proposed as sensitive and potential early-warning tools. In recent years, interactions between different chemicals (mixture toxicity) and between chemical and other stressors attracted scientific interest. With the development of molecular biology, omics tools are gaining increasing interest, while the ecological relevance of exposure and effects is translating into concepts like (chemical) stress ecology, ecological vulnerability and trait-based approaches. This contribution addresses historical developments and focuses on current issues in soil ecotoxicology. It is concluded that soil ecotoxicological risk assessment would benefit from extending the available battery of toxicity tests by including e.g. isopods, by paying more attention to exposure, bioavailability and toxicokinetics, and by developing more insight into the ecology of soil organisms to support better understanding of exposure and long-term consequences of chemical exposure at the individual, population and community level. Ecotoxicogenomics tools may also be helpful in this, but will require considerable further research before they can be applied in the practice of soil ecotoxicological risk assessment.

Ecotoxicological Risk Assessment

In ecotoxicological risk assessment, two approaches can be distinguished. One approach aims at predicting possible effects of (new) chemicals in order to regulate their use or prevent their introduction onto the market. This predictive approach (*prognosis*) uses laboratory toxicity tests to obtain toxicity data that are used to derive safe levels of chemicals in the environment. The second approach is assessing the actual ecological risk or damage in case of pollution. This diagnostic approach (*diagnosis*) enables setting priorities for remediation and risk reduction, and may provide triggers for the management of contaminated land.

Prognosis starts from the paradigms also used in human toxicology. It assumes that the risk of a chemical for ecosystems can be estimated from its toxicity to a number of surrogate test or indicator species, exposed in standard laboratory toxicity tests. These tests aim at assessing toxicity, which is expressed in terms of dose-response relationships for effects on selected endpoints like survival, growth and reproduction. Toxicity is quantified by parameters like LC_{10} and LC_{50} (the concentrations killing 10% and 50% of the exposed test organisms, respectively), EC_{10} and EC_{50} (the concentrations causing 10% and 50% reduction, respectively in a measured endpoint, e.g. growth or number of juveniles produced), and NOEC and LOEC (no-observable and lowest-observable effect concentration, respectively). Since there is no 'most sensitive species' a battery of toxicity tests is needed to obtain proper insight into the potential hazard of a chemical for the ecosystem. In prognosis, the outcome of toxicity tests is used to establish thresholds or safe levels of chemicals in soil, which can be compared with measured or predicted exposure data (soil concentrations) to assess the (potential) risk.

A critical part of this procedure is the derivation of safe levels of chemicals on the basis of available toxicity data. When only short-term (acute) toxicity data are available (usually focusing on survival) or data for a limited number of species, somewhat arbitrary application factors are applied to derive safe levels that should protect ecosystems. For example, when only one or two LC_{50} values are available, a factor of 1000 is applied to the lowest LC_{50} value; this factor should be sufficient to extrapolate from acute to chronic effects (factor of 10), from one or few species to many species (factor of 10), and from laboratory to field (factor of 10). When sublethal toxicity data (usually NOEC or EC_{10} values for effects on e.g. reproduction) are available for 3 or more species, application of a factor of 10 to the lowest value is considered sufficiently protective. When many toxicity data are available (preferably ≥ 8) for species representative of different taxonomic groups, a statistical method may be applied. Such a statistical method is used to construct a species-sensitivity distribution (SSD), which assumes a log-normal or log-logistic distribution of the sensitivities of species in an ecosystem. From such an SSD the 95% lower limit is selected as the safe level. At this level, at least 95% of the species in the ecosystem are supposed to be safe.

Diagnosis may use the same tools as applied for prognoses, but it more heavily relies on ecological methods and environmental chemistry. Basically, toxicity tests or bioassays are used as diagnosis tools to assess toxicity of soil samples from a contaminated site. Results of the bioassays, together with those of chemical measurements and ecological field observations, are used to assess the potential risk of soil contamination. The three tools together form the TRIAD approach.

Toxicity Tests

Both prognosis and diagnosis use toxicity tests, and in both cases a battery of tests is recommended. Criteria to select tests for such a battery have been formulated e.g. by Van Gestel These criteria among others include:

- Practicability, referring to the feasibility and cost-effectiveness of a test.
- Acceptability, including aspects like standardization, reproducibility and statistical validity of a test method as well as its broad chemical responsiveness.
- Ecological meaning, including sensitivity and ecological realism of the test method.

To obtain a balanced battery of tests, in addition the following criteria need to be taken into account:

- Representativeness for the ecosystem to protect: this includes e.g. the representation of organisms having different life-histories, representing different functional groups, different taxonomic groups and different routes of exposure.

- Representativeness of responses, to make sure responses measured really are relevant for the protection of populations and communities.
- Uniformity, which refers to the possibility to apply all tests in a battery to the same test media.

By the end of the 1990s and early 2000 toxicity tests, using sub-lethal endpoints like reproduction, were standardized for enchytraeids, earthworms and Collembola by both the OECD and the International Standardization Organization (ISO). But also Environment Canada, the United States Environmental Protection Agency (EPA) and ASTM International (formerly known as the American Society for Testing and Materials) have described similar methods. Recently, for the same organisms, avoidance behavior tests have been described, while for earthworms and enchytraeids a bioaccumulation test is available.

The oldest standardized toxicity test guideline with soil invertebrates, OECD 207, describes two short-term toxicity tests, one using 14 days exposure in soil and the other one exposing the worms for 2 days to treated filter paper. Both methods use survival as the only endpoint. The test on filter paper is only rarely applied nowadays, as it does not have any relevance for exposure in soil. It may however, be a useful exposure method for the rapid screening of chemicals, assessing the uptake and biotransformation of chemicals or other types of mechanistic research.

Compared to survival, reproduction is a more relevant endpoint when translating effects to the population level. For that reason, an earthworm toxicity test focusing on reproduction has been developed. Although focus is on reproduction, it is essential in this test to also include weight change of the earthworms, since there is evidence for a trade off between reproduction and growth, which may affect their response to toxicants. Like for earthworms, the tests with enchytraeids also focus on reproduction and survival as the endpoint, but for these organisms no separate short-term (acute) and sub-lethal tests were developed. All the reproduction toxicity tests available with oligochaetes typically have test durations of 21-28 days. In case of the earthworm test with *Eisenia fetida* or *Eisenia andrei* and the enchytraeid test with *Enchytraeus albidus*, this means exposing adult worms for 28 and 21 days, respectively, after which they are collected from the soil; the cocoons are incubated for another 28 or 21 days, respectively to enable determining the number of juveniles produced. Nowadays, *Enchytraeus crypticus* seems to be more commonly used for toxicity testing than *Enchytraeus albidus*. Since that species is smaller, adults are not removed from the soil and reproduction is determined after 28 days of exposure. Considering the shorter life cycle of *Enchytraeus crypticus* and its high reproductive output, limiting test duration to 21 days has recently been advocated.

A standardized test with snails has been developed by ISO, focusing on survival and growth of juveniles snails (*Helix aspersa aspersa*) after 28 days exposure.

Standardized toxicity tests with soil arthropods include the collembolan species *Folsomia candida* and *Folsomia fimetaria*, the predatory mite *Hypoaspis (Geolaelaps) aculeifer*, and larvae of the insect *Oxythyrea funesta*. The tests with the collembolans and the predatory mite typically focus on survival and reproduction after 28, 21 and 14 days exposure, respectively for the parthenogenetic *Folsomia candida*, the sexually reproducing *Folsomia fimetaria* and the predatory mite *Hypoaspis aculeifer*. The test with insect larvae only focuses on survival.

Tests with carabid beetles have been performed using adult *Pterostichus oblongopunctatus* or larvae of *Poecilus cupreus*, but these tests have not been standardized and use different life stages (larvae, adults), endpoints (survival, mobility, respiration) and test durations (from few weeks to several months) depending on the aims of the study.

Also the toxicity tests with the oribatid mites *Oppia nitens* and *Platynothrus peltifer* described in the literature are not yet standardized. These tests focus on survival and reproduction.

For assessing chemical toxicity to isopods, also no standard test guidelines are available. Nevertheless, isopods are used as test organisms, using different test durations, different routes of exposure (food, soil) and different endpoints. Drobne and Hopkin described a test exposing *Porcellio scaber* via food and determining effects of zinc on feeding rates. Hornung developed a draft test guideline, allowing for exposure both via food and in soil, also using *Porcellio scaber* as the test species. Several different endpoints have been proposed, but considering the difficulty of culturing *Porcellio scaber* in the laboratory survival, growth and food consumption rate so far have been used most frequently. Another interesting endpoint may be related to the composition of the gut microflora of isopods. Nevertheless, isopod (*Porcellio scaber*) reproduction also has been proposed as a test endpoint. Successful reproduction isopod experiments have been performed, e.g. by Van Brummelen, in a 48-week test with *Oniscus asellus*, applying dietary exposure to Polycyclic Aromatic Hydrocarbons (PAHs), and by Lemos assessing the reproduction toxicity to *Porcellio scaber* of some chemicals with suspected endocrine-disrupting effects. Other isopod species, like *Porcellionides pruinosis*, are more easily cultured in the laboratory, and therefore may be more suitable for performing reproduction toxicity tests.

Recently, avoidance response was introduced as an easy, fast and sensitive endpoint. For some chemicals avoidance response may be as sensitive as reproduction, while for others it is at least as sensitive as survival. Great advantage of avoidance tests is that they are fast, with test durations of no more than 2 days. Standard test guidelines for avoidance tests have been developed for earthworms and Collembola, but similar tests have also been performed with enchytraeids, oribatid mites and isopods.

In addition to these tests with soil invertebrates, ISO and OECD have also developed a number of toxicity tests with plants, which are important in soil ecosystems as primary producers. Also, several tests are available focusing on the effects of microbial communities or processes performed by microorganisms, like nitrification.

Considering the fact that for a proper risk assessment a battery of tests is desirable, it is important to consider the currently available test methods. The current set of available tests shows an underrepresentation of arthropods in comparison with their abundance in the field when compared with other species like Oligochaetes. And of the available or suggested tests with arthropods, only the one with Collembola has been standardized. Development and international standardization of more toxicity tests with representative arthropod species therefore is highly needed. The ecological relevance of isopods, their typical routes of exposure (soil, food) and life history characteristics, the possibility to determine different endpoint, and the fact that they have already been used for testing for more than 30 years, make them highly suitable test organisms. Standardization of toxicity tests with isopods therefore is highly recommended.

Bioavailability

For reasons of standardization and to facilitate comparison of results, all standardized tests use a standard soil type: the so-called OECD artificial soil, first introduced in the earthworm acute toxicity test developed by OECD. This artificial soil is composed by mixing readily available materials like sphagnum peat (10%), kaolin clay (20%) and quartz sand (70%); by adding some $CaCO_3$ pH is adjusted to approx. 6.0. The properties of this soil resemble those of a sandy loam soil. Recently, within OECD the use of 5% peat has been advocated when testing pesticides, in order to increase the 'worst case' realism of the artificial test soil for low organic (agricultural) field soils. Also with the aim to increase realism, the SECOFASE project started using the natural LUFA 2.2 soil. The LUFA soils are commercially available from the Landwirtschaftliche Untersuchungs und Forschungsanstalt (LUFA) in Speyer, Germany. Since that time, LUFA 2.2 standard soil seems to become more commonly used for toxicity tests with soil invertebrates, also because several test species like the collembolan *Folsomia candida* and the enchytraeid *Enchytraeus crypticus* seem to perform as good in this natural soil as they do in artificial soil.

The notion that soil type was important when determining the toxicity of chemicals went along with the increasing insight into the concept of bioavailability: only a fraction of the total amount of chemical in the soil is available for uptake by organisms and therefore of relevance for risk assessment. This was, for instance, demonstrated by Bradham, exposing the earthworm *Eisenia andrei* for 28 days to different soil types spiked with one concentration of lead (Pb). While in some soils all worms died, in other soils no mortality was seen and in the remainder of the soils only part of the earthworms died. This finding could be explained from the differences in soil properties, especially pH, clay and organic matter content that affected the availability of lead for the earthworms. The pore-water hypothesis or equilibrium partitioning theory was developed to enable linking toxicity of organic chemicals to the concentration available in the pore water. For metals, application of this approach turned out to be less straightforward because metal speciation in soils is much more complex and many factors may affect metal bioavailability in soils. It nowadays is realized that bioavailability is not a static but rather a dynamic concept. This was demonstrated by Van Straalen, exposing isopods (*Porcellio scaber*) from a metal-contaminated and a non-polluted site to zinc via the food. Both groups showed the same EC_{50} expressed on the basis of zinc concentrations in the diet, but contrary to the expectation effects could not be explained from zinc concentrations in the animals. So, it seems that flux of zinc through the animals rather than total zinc concentration was determining their sensitivity.

These findings also suggest that when considering bioavailability, not only chemical partitioning of chemicals in the exposure medium (soil, food) and pore-water concentrations have to be considered. The biology of bioavailability also needs attention. One aspect of this is the way organisms deal with chemicals. For metals, internal compartmentalization has been shown to be an important aspect, as it determines what fraction of the metal is present in the body in a metabolically available form. Most soil invertebrates have the capacity to sequester at least part of their metal burden in such a way that it does no longer pose a risk. Isopods use the hepatopancreas for a very efficient storage of excess metals, while earthworms have their chloragogenous tissue that serves the same purpose. As a consequence, both isopods and earthworms show a huge capacity of storing metals like cadmium, which after uptake are hardly eliminated. Other soil invertebrates, like Collembola and beetles, use the midgut epithelium for metal storage. Upon moulting, also the midgut epithelium is renewed enabling these organisms to excrete excess metal. Internal sequestration determines what fraction of the total metal burden in an organism may contribute to its toxicity or is available for trophic transfer to its predators. This was for instance demonstrated by Crommentuijn, who found very high Lethal Body Concentrations for cadmium in isopods compared to other arthropods that could be attributed to the highly efficient storage of the metal in an inert form.

Another way organisms may deal with potentially toxic chemicals is by biotransformation. The process of biotransformation aims at making chemicals more hydrophilic and in this way facilitating their excretion. Isopods and Collembola have been shown to be extremely efficient in biotransforming organic chemicals like Polycyclic Aromatic Hydrocarbons (PAHs), which are excreted by these organisms with half lives of approximately 1 day. Earthworms seem less efficient in doing so. Possible consequence of this rapid biotransformation is that potentially toxic metabolites may be produced. This has been shown in isopods, with even DNA adducts being formed upon exposure to PAHs. This may also lead to long-term damage and possible effects on subsequent generations as has been shown for phenanthrene in the collembolan *Folsomia candida*. As very little is known about such multi-generation effects, further research on the long-term effects of chemicals is urgently needed.

An important biological aspect that may affect the exposure of soil invertebrates to chemicals is their behavior. Soil by definition is a heterogeneous environment. As a consequence, also the distribution of chemicals in soil is heterogeneous. Chemicals reaching soil by areal deposition for instance accumulate in the topsoil layer, leading to a depth-related concentration gradient as was shown for PAHs in forest soils. Depending on the habitat and mobility, organisms may be more or less exposed to chemicals present in the topsoil layer. Similarly, the effect of pesticides on earthworms was shown to be highly related to their mobility with epigeic and anecic species being much more vulnerable compared to endogeic species. Especially anecic species, like *Lumbricus terrestris*, which come to the soil surface to forage and mate at night, may experience a very high exposure shortly after pesticide spraying. Also in case of spatially heterogeneous soil pollution, behavior may affect exposure, as was shown for earthworm exposure to copper in a heterogeneously polluted soil by Marinussen and Van der Zee knowledge of the uptake and elimination kinetics showed to be very helpful in predicting metal concentrations in the earthworms living in a heterogeneously polluted environment. Also in case of isopods, behavior is an important factor determining exposure. Unfortunately, no research has been done on the chemical exposure of isopods and consequent effects in relation to their behavior in the field.

Multiple Species, Model Ecosystem (Microcosm) and Field Tests

All standardized toxicity tests with soil invertebrates focus on assessing the effects of chemicals on single species of organisms. To enable assessment of toxic effects in a more realistic setting, micro- or model ecosystems have been developed, ranging from artificially composed set-ups with a number of selected different species introduced in a well homogenized soil to intact soil columns extracted from the field and incubated in the laboratory. Such model ecosystems or microcosms allow assessing effects at the community level, taking into account the interactions between species. Although basically considered single-species tests, earthworm and isopod tests focusing on decomposition or feeding activity in fact also are multispecies tests as in these tests also the interaction with microorganisms in the gut and in the soil or food are important. In addition, the endpoint (decomposition) has high ecological relevance for assessing potential effects on the functioning of the soil ecosystem. The only field test available aims at assessing pesticide effects on earthworms, but can be combined with a decomposition or litter bag test.

Since the introduction of the term Ecotoxicology, the question for "putting more eco into ecotoxicology" has been raised. Some authors even argued that ecotoxicology should not be seen as a sub-discipline of toxicology but rather as a case of *stress ecology*. This notion has triggered the focus on more ecologically relevant test designs, integrated approaches including responses at different levels of biological organization, and taking into account the normal operating range of parameters describing the structure and functioning of soil ecosystems.

Since early 2000, with the notion of stress ecology, more complex issues have been receiving attention, with ecological vulnerability, trait-based analysis and effects on functional endpoints (so-called *ecosystem services*) being key items. The application of these trait-based approaches in soil ecotoxicology on one hand offers promising perspectives, on the other hand it also demonstrates an enormous lack of knowledge on the traits represented by different species and groups of soil invertebrates.

Persistent Organic Pollutant

Persistent organic pollutants (POPs), sometimes known as "forever chemicals" are organic compounds that are resistant to environmental degradation through chemical, biological, and photolytic processes. Because of their persistence, POPs bioaccumulate with potential adverse impacts on human health and the environment. The effect of POPs on human and environmental health was discussed, with intention to eliminate or severely restrict their production, by the international community at the Stockholm Convention on Persistent Organic Pollutants in 2001.

Many POPs are currently or were in the past used as pesticides, solvents, pharmaceuticals, and industrial chemicals. Although some POPs arise naturally, for example volcanoes and various biosynthetic pathways, most are man-made via total synthesis.

Consequences of Persistence

POPs typically are halogenated organic compounds and as such exhibit high lipid solubility. For this reason, they bioaccumulate in fatty tissues. Halogenated compounds also exhibit great stability reflecting the nonreactivity of C-Cl bonds toward hydrolysis and photolytic degradation. The stability and lipophilicity of organic compounds often correlates with their halogen content, thus polyhalogenated organic compounds are of particular concern. They exert their negative effects on the environment through two processes, long range transport, which allows them to travel far from their source, and bioaccumulation, which reconcentrates these chemical compounds to potentially dangerous levels. Compounds that make up POPs are also classed as PBTs (Persistent, Bioaccumulative and Toxic) or TOMPs (Toxic Organic Micro Pollutants).

Long-range Transport

POPs enter the gas phase under certain environmental temperatures and volatize from soils, vegetation, and bodies of water into the atmosphere, resisting breakdown reactions in the air, to travel long distances before being re-deposited. This results in accumulation of POPs in areas far from where they were used or emitted, specifically environments where POPs have never been introduced such as Antarctica, and the Arctic circle. POPs can be present as vapors in the atmosphere or bound to the surface of solid particles. POPs have low solubility in water but are easily captured by solid particles, and are soluble in organic fluids (oils, fats, and liquid fuels). POPs are not easily degraded in the environment due to their stability and low decomposition rates. Due to this capacity for long-range transport, POP environmental contamination is extensive, even in areas where POPs have never been used, and will remain in these environments years after restrictions implemented due to their resistance to degradation.

Bioaccumulation

Bioaccumulation of POPs is typically associated with the compounds high lipid solubility and ability to accumulate in the fatty tissues of living organisms for long periods of time. Persistent chemicals tend to have higher concentrations and are eliminated more slowly. Dietary accumulation or bioaccumulation is another hallmark characteristic of POPs, as POPs move up the food chain, they increase in concentration as they are processed and metabolized in certain tissues of organisms. The natural capacity for animals gastrointestinal tract concentrate ingested chemicals, along with poorly metabolized and hydrophobic nature of POPs makes such compounds highly susceptible to bioaccumulation. Thus POPs not only persist in the environment, but also as they are taken in by animals they bioaccumulate, increasing their concentration and toxicity in the environment.

Stockholm Convention on Persistent Organic Pollutants

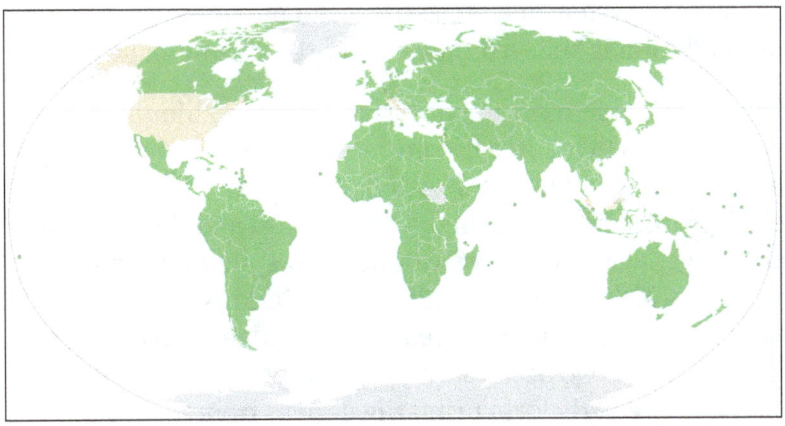

State parties to the Stockholm Convention on Persistent Organic Pollutants.

The Stockholm Convention was adopted and put into practice by the United Nations Environment Programme (UNEP) on May 22, 2001. The UNEP decided that POP regulation needed to be addressed globally for the future. The purpose statement of the agreement is "to protect human health and the environment from persistent organic pollutants." As of 2014, there are 179 countries in compliance with the Stockholm convention. The convention and its participants have recognized the potential human and environmental toxicity of POPs. They recognize that POPs have the potential for long range transport and bioaccumulation and biomagnification. The convention seeks to study and then judge whether or not a number of chemicals that have been developed with advances in technology and science can be categorized as POPs or not. The initial meeting in 2001 made a preliminary list, termed the "dirty dozen," of chemicals that are classified as POPs. As of 2014, the United States of America has signed the Stockholm Convention but has not ratified it. There are a handful of other countries that have not ratified the convention but most countries in the world have ratified the convention.

Compounds on the Stockholm Convention List

In May 1995, the United Nations Environment Programme Governing Council investigated POPs. Initially the Convention recognized only twelve POPs for their adverse effects on human health and the environment, placing a global ban on these particularly harmful and toxic compounds and requiring its parties to take measures to eliminate or reduce the release of POPs in the environment.

- Aldrin, an insecticide used in soils to kill termites, grasshoppers, Western corn rootworm, and others, is also known to kill birds, fish, and humans. Humans are primarily exposed to aldrin through dairy products and animal meats.
- Chlordane, an insecticide used to control termites and on a range of agricultural crops, is known to be lethal in various species of birds, including mallard ducks, bobwhite quail, and pink shrimp; it is a chemical that remains in the soil with a reported half-life of one year. Chlordane has been postulated to affect the human immune system and is classified as a possible human carcinogen. Chlordane air pollution is believed the primary route of humane exposure.
- Dieldrin, a pesticide used to control termites, textile pests, insect-borne diseases and insects living in agricultural soils. In soil and insects, aldrin can be oxidized, resulting in rapid conversion to dieldrin. Dieldrin's half-life is approximately five years. Dieldrin is highly toxic to fish and other aquatic animals, particularly frogs, whose embryos can develop spinal deformities after exposure to low levels. Dieldrin has been linked to Parkinson's disease, breast cancer, and classified as immunotoxic, neurotoxic, with endocrine disrupting capacity. Dieldrin residues have been found in air, water, soil, fish, birds, and mammals. Human exposure to dieldrin primarily derives from food.
- Endrin, an insecticide sprayed on the leaves of crops, and used to control rodents. Animals can metabolize endrin, so fatty tissue accumulation is not an issue, however the chemical has a long half-life in soil for up to 12 years. Endrin is highly toxic to aquatic animals and humans as a neurotoxin. Human exposure results primarily through food.
- Heptachlor, a pesticide primarily used to kill soil insects and termites, along with cotton insects, grasshoppers, other crop pests, and malaria-carrying mosquitoes. Heptachlor, even at every low doses has been associated with the decline of several wild bird populations – Canada geese and American kestrels. In laboratory tests have shown high-dose heptachlor as lethal, with adverse

behavioral changes and reduced reproductive success at low-doses, and is classified as a possible human carcinogen. Human exposure primarily results from food.

- Hexachlorobenzene (HCB), was first introduced in 1945–59 to treat seeds because it can kill fungi on food crops. HCB-treated seed grain consumption is associated with photosensitive skin lesions, colic, debilitation, and a metabolic disorder called porphyria turcica, which can be lethal. Mothers who pass HCB to their infants through the placenta and breast milk had limited reproductive success including infant death. Human exposure is primarily from food.

- Mirex, an insecticide used against ants and termites or as a flame retardant in plastics, rubber, and electrical goods. Mirex is one of the most stable and persistent pesticides, with a half-life of up to 10 years. Mirex is toxic to several plant, fish and crustacean species, with suggested carcinogenic capacity in humans. Humans are exposed primarily through animal meat, fish, and wild game.

- Toxaphene, an insecticide used on cotton, cereal, grain, fruits, nuts, and vegetables, as well as for tick and mite control in livestock. Widespread toxaphene use in the US and chemical persistence, with a half-life of up to 12 years in soil, results in residual toxaphene in the environment. Toxaphene is highly toxic to fish, inducing dramatic weight loss and reduced egg viability. Human exposure primarily results from food. While human toxicity to direct toxaphene exposure is low, the compound is classified as a possible human carcinogen.

- Polychlorinated biphenyls (PCBs), used as heat exchange fluids, in electrical transformers, and capacitors, and as additives in paint, carbonless copy paper, and plastics. Persistence varies with degree of halogenation, an estimated half-life of 10 years. PCBs are toxic to fish at high doses, and associated with spawning failure at low doses. Human exposure occurs through food, and is associated with reproductive failure and immune suppression. Immediate effects of PCB exposure include pigmentation of nails and mucous membranes and swelling of the eyelids, along with fatigue, nausea, and vomiting. Effects are transgenerational, as the chemical can persist in a mother's body for up to 7 years, resulting in developmental delays and behavioral problems in her children. Food contamination has led to large scale PCB exposure.

- Dichlorodiphenyltrichloroethane (DDT) is probably the most infamous POP. It was widely used as insecticide during WWII to protect against malaria and typhus. After the war, DDT was used as an agricultural insecticide. In 1962, the American biologist Rachel Carson published Silent Spring, describing the impact of DDT spraying on the US environment and human health. DDT's persistence in the soil for up to 10–15 years after application has resulted in widespread and persistent DDT residues throughout the world including the arctic, even though it has been banned or severely restricted in most of the world. DDT is toxic to many organisms including birds where it is detrimental to reproduction due to eggshell thinning. DDT can be detected in foods from all over the world and food-borne DDT remains the greatest source of human exposure. Short-term acute effects of DDT on humans are limited, however long-term exposure has been associated with chronic health effects including increased risk of cancer and diabetes, reduced reproductive success, and neurological disease.

- Dioxins are unintentional by-products of high-temperature processes, such as incomplete combustion and pesticide production. Dioxins are typically emitted from the burning of hospital waste, municipal waste, and hazardous waste, along with automobile emissions, peat, coal, and wood. Dioxins have been associated with several adverse effects in humans, including immune and enzyme disorders, chloracne, and are classified as a possible human carcinogen. In laboratory studies of dioxin effects an increase in birth defects and stillbirths, and lethal exposure have been associated with the substances. Food, particularly from animals, is the principal source of human exposure to dioxins.
- Polychlorinated dibenzofurans are by-products of high-temperature processes, such as incomplete combustion after waste incineration or in automobiles, pesticide production, and polychlorinated biphenyl production. Structurally similar to dioxins, the two compounds share toxic effects. Furans persist in the environment and classified as possible human carcinogens. Human exposure to furans primarily results from food, particularly animal products.

New POPs on the Stockholm Convention List

Since 2001, this list has been expanded to include some polycyclic aromatic hydrocarbons (PAHs), brominated flame retardants, and other compounds. Additions to the initial 2001 Stockholm Convention list are as following POPs:

- Chlordecone, a synthetic chlorinated organic compound, is primarily used as an agricultural pesticide, related to DDT and Mirex. Chlordecone is toxic to aquatic organisms, and classified as a possible human carcinogen. Many countries have banned chlordecone sale and use, or intend to phase out stockpiles and wastes.
- α-Hexachlorocyclohexane (α-HCH) and β-Hexachlorocyclohexane (β-HCH) are insecticides as well as by-products in the production of lindane. Large stockpiles of HCH isomers exist in the environment. α-HCH and β-HCH are highly persistent in the water of colder regions. α-HCH and β-HCH has been linked Parkinson's and Alzheimer's disease.
- Hexabromodiphenyl ether (hexaBDE) and heptabromodiphenyl ether (heptaBDE) are main components of commercial octabromodiphenyl ether (octaBDE). Commercial octaBDE is highly persistent in the environment, whose only

degradation pathway is through debromination and the production of bromodiphenyl ethers, which can increase toxicity.

- Lindane (γ-hexachlorocyclohexane), a pesticide used as a broad spectrum insecticide for seed, soil, leaf, tree and wood treatment, and against ectoparasites in animals and humans (head lice and scabies). Lindane rapidly bioconcentrates. It is immunotoxic, neurotoxic, carcinogenic, linked to liver and kidney damage as well as adverse reproductive and developmental effects in laboratory animals and aquatic organisms. Production of lindane unintentionally produces two other POPs α-HCH and β-HCH.

- Pentachlorobenzene (PeCB), is a pesticide and unintentional byproduct. PeCB has also been used in PCB products, dyestuff carriers, as a fungicide, a flame retardant, and a chemical intermediate. PeCB is moderately toxic to humane, while highly toxic to aquatic organisms.

- Tetrabromodiphenyl ether (tetraBDE) and pentabromodiphenyl ether (pentaBDE) are industrial chemicals and the main components of commercial pentabromodiphenyl ether (pentaBDE). PentaBDE has been detected in humans in all regions of the world.

- Perfluorooctanesulfonic acid (PFOS) and its salts are used in the production of fluoropolymers. PFOS and related compounds are extremely persistent, bioaccumulating and biomagnifying. The negative effects of trace levels of PFOS have not been established.

- Endosulfans are insecticides to control pests on crops such coffee, cotton, rice and sorghum and soybeans, tsetse flies, ectoparasites of cattle. They are used as a wood preservative. Global use and manufacturing of endosulfan has been banned under the Stockholm convention in 2011, although many countries had previously banned or introduced phase-outs of the chemical when the ban was announced. Toxic to humans and aquatic and terrestrial organisms, linked to congenital physical disorders, mental retardation, and death. Endosulfans' negative health effects are primarily liked to its endocrine disrupting capacity acting as an antiandrogen.

- Hexabromocyclododecane (HBCD) is a brominated flame retardant primarily used in thermal insulation in the building industry. HBCD is persistent, toxic and ecotoxic, with bioaccumulative and long-range transport properties.

Health Effects

POP exposure may cause developmental defects, chronic illnesses, and death. Some are carcinogens per IARC, possibly including breast cancer. Many POPs are capable of endocrine disruption within the reproductive system, the central nervous system, or the immune system. People and animals are exposed to POPs mostly through their diet, occupationally, or while growing in the womb. For humans not exposed to POPs through accidental or occupational means, over 90% of exposure comes from animal product foods due to bioaccumulation in fat tissues and bioaccumulate through the food chain. In general, POP serum levels increase with age and tend to be higher in females than males.

Studies have investigated the correlation between low level exposure of POPs and various diseases. In order to assess disease risk due to POPs in a particular location, government agencies may produce a human health risk assessment which takes into account the pollutants' bioavailability and their dose-response relationships.

Endocrine Disruption

The majority of POPs are known to disrupt normal functioning of the endocrine system. Low level exposure to POPs during critical developmental periods of fetus, newborn and child can have a lasting effect throughout their lifespan. A 2002 study summarizes data on endocrine disruption and health complications from exposure to POPs during critical developmental stages in an organism's lifespan. The study aimed to answer the question whether or not chronic, low level exposure to POPs can have a health impact on the endocrine system and development of organisms from different species. The study found that exposure of POPs during a critical developmental time frame can produce a permanent changes in the organisms path of development. Exposure of POPs during non-critical developmental time frames may not lead to detectable diseases and health complications later in their life. In wildlife, the critical development time frames are in utero, in ovo, and during reproductive periods. In humans, the critical development timeframe is during fetal development.

Reproductive System

The same study in 2002 with evidence of a link from POPs to endocrine disruption also linked low dose exposure of POPs to reproductive health effects. The study stated that POP exposure can lead to negative health effects especially in the male reproductive system, such as decreased sperm quality and quantity, altered sex ratio and early puberty onset. For females exposed to POPs, altered reproductive tissues and pregnancy outcomes as well as endometriosis have been reported.

Gestational Weight Gain and Newborn Head Circumference

The maternal weight gain during pregnancy, their PCB-exposure level and PCB level in their newborn infants, their birth weight, gestational age, and head circumference. The lower the birth weight and head circumference of the infants was, the higher POP levels during prenatal development had been, but only if mothers had either excessive or inadequate weight gain during pregnancy. No correlation between POP exposure and gestational age was found. A 2013 case-control study conducted 2009 in Indian mothers and their offspring showed prenatal exposure of two types of organochlorine pesticides (HCH, DDT and DDE) impaired the growth of the fetus, reduced the birth weight, length, head circumference and chest circumference.

Additive and Synergistic Effects

Evaluation of the effects of POPs on health is very challenging in the laboratory setting. For example, for organisms exposed to a mixture of POPs, the effects are assumed to be additive. Mixtures of POPs can in principle produce synergistic effects. With synergistic effects, the toxicity of each compound is enhanced (or depressed) by the presence of other compounds in the mixture. When put together, the effects can far exceed the approximated additive effects of the POP compound mixture.

Urban Areas and Indoor Environments

Traditionally it was thought that human exposure to POPs occurred primarily through food, however indoor pollution patterns that characterize certain POPs have challenged this notion. Recent studies of indoor dust and air have implicated indoor environments as a major sources for human exposure via inhalation and ingestion. Furthermore, significant indoor POP pollution must be a major route of human POP exposure, considering the modern trend in spending larger proportions of life indoor. Several studies have shown that indoor (air and dust) POP levels to exceed outdoor (air and soil) POP concentrations.

Control and Removal in the Environment

Current studies aimed at minimizing POPs in the environment are investigating their behavior in photo catalytic oxidation reactions. POPs that are found in humans and in aquatic environments the most are the main subjects of these experiments. Aromatic and aliphatic degradation products have been identified in these reactions. Photochemical degradation is negligible compared to photocatalytic degradation. A method of removal of POPs from marine environments that has been explored is adsorption. It occurs when an absorbable solute comes into contact with a solid with a porous surface structure. This technique was investigated by Mohamed Nageeb Rashed of Aswan University, Egypt. Current efforts are more focused on banning the use and production of POPs worldwide rather than removal of POPs.

Persistent, Bioaccumulative and Toxic Substances

Persistent, bioaccumulative and toxic substances (PBTs) are a class of compounds that have high resistance to degradation from abiotic and biotic factors, high mobility in the environment and high toxicity. Because of these factors PBTs have been observed to have a high order of bioaccumulation and biomagnification, very long retention times in various media, and widespread distribution across the globe. Majority of PBTs in the environment are either created through industry or are unintentional byproducts.

Persistent organic pollutants (POPs) were the focal point of the Stockholm Convention 2001 due to their persistence, ability to biomagnify and the threat posed to both human health and the environment. The goal of the Stockholm Convention was to determine the classification of POPs, create measures to eliminate production/use of POPs, and establish proper disposal of the compounds in an environmentally friendly manner. Currently the majority of the global community is actively involved with this program but a few still resist, most notably the U.S.

Similar to POPs classification, the PBT classification of chemicals was developed in 1997 by the Great Lakes Binational Toxic Strategy (GLBNS). Signed by both the U.S and Canada, the GLBNS classified PBTs in one of two categories, level I and level II. Level I PBTs are top priority which currently, as of 2005, contained 12 compounds or classes of compounds.

Level I PBTs (GLBNS)

- Mercury.
- Polychlorinated biphenyls PCBs).
- Dioxins/furans.
- Benzo(a)pyrene (BaP).
- Hexachlorobenzene (HCB).
- Alkyl-lead.
- Pesticides.
    - Mirex.
    - Dieldrin/aldrin.
    - Chlordane.
    - Toxaphene.
- Octachlorostyrene.

The GLBNS is administered by the U.S Environmental Protection Agency (USEPA) and Environment Canada. Following the GLBNS, the Multimedia Strategy for Priority Persistent, Bioaccumulative and Toxic Pollutants (PBT Strategy) was drafted by the USEPA. The PBT Strategy led to the implementation of PBT criteria in several regulational policies. Two main policies that were changed by the PBT strategy were the Toxics Release Inventory (TRI) which required more rigid chemical reporting and the New Chemical Program (NCP) under the Toxics Substances Control Act (TSCA) which required screening for PBTs and PBT properties.

Compounds

General

PBTs are a unique classification of chemicals that have and will continue to impact human health and the environment worldwide. The three main attributes of a PBT (persistence, bioaccumulative and toxic) each have a huge role in the risk posed by these compounds.

Persistence

PBTs may have a high environmental mobility relative to other contaminants mainly due to their resistance to degradation (persistence). This allows PBTs to travel far and wide in both the atmosphere and in aqueous environments. The low degradation rates of PBTs allow these chemicals to be exposed to both biotic and abiotic factors while maintaining a relatively stable concentration. Another factor that makes PBTs especially dangerous are the degradation products which are often relatively as toxic as the parent compound. These factors have resulted in global contamination most notable in remote areas such as the arctic and high elevation areas which are far from any source of PBTs.

Bioaccumulation and Biomagnification

The bioaccumulative ability of PBTs follows suit with the persistence attribute by the high resistance to degradation by biotic factors, especially with in organisms. Bioaccumulation is the result of a toxic substance being taken up at a higher rate than being removed from an organism. For PBTs this is caused mainly by a resistance to degradation, biotic and abiotic. PBTs usually are highly insoluble in water which allows them to enter organisms at faster rates through fats and other nonpolar regions on an organism. Bioaccumulation of a toxicant can lead to biomagnification through a trophic web which has resulted in massive concern in areas with especially low trophic diversity. Biomagnification results in higher trophic organisms accumulating more PBTs than those of lower trophic levels through consumption of the PBT contaminated lower trophic organisms.

Toxicity

The toxicity of this class of compounds is high with very low concentrations of a PBT required to enact an effect on an organism compared to most other contaminants. This high toxicity along with the persistence allows for the PBT to have detrimental effects in remote areas around the globe where there is not a local source of PBTs. The bioaccumulation and magnification along with the high toxicity and persistence has the ability to destroy and irreparably damage trophic systems, especially the higher trophic levels, globally. It is this reason that PBTs have become an area of focus in global politics.

Specific Toxicants

PCBs

Historically, PCBs were used extensively for industrial purposes such as coolants, insulating fluids, and as a plasticizer. These contaminants enter the environment through both use and disposal. Due to extensive concern from the public, legal, and scientific sectors indicating that PCBs are likely carcinogens and potential to adversely impact the environment, these compounds were banned in 1979 in the United States. The ban included the use of PCBs in uncontained sources, such as adhesives, fire retardant fabric treatments, and plasticizers in paints and cements. Containers that are completely enclosed such as transformers and capacitors are exempt from the ban.

The inclusion of PCBs as a PBT may be attributed to their low water solubility, high stability, and semi-volatility facilitating their long range transport and accumulation in organisms. The persistence of these compounds is due to the high resistance to oxidation, reduction, addition, elimination and electrophilic substitution. The toxicological interactions of PCBs are affected by the number and position of the chlorine atoms, without ortho substitution are referred as coplanar and all others as non-coplanar. Non-coplanar PCBs may cause neurotoxicity by interfering with intracellular signal transduction dependent on calcium. Ortho-PCBs may alter hormone regulation through disruption of the thyroid hormone transport by binding to transthyretin. Coplanar PCBs are similar to dioxins and furans, both bind to the aryl hydrocarbon receptor (AhR) in organisms and may exert dioxin-like effects, in addition to the effects shared with non-coplanar PCBs. The AhR is a transcription factor, therefore, abnormal activation may disrupt cellular function by altering gene transcription.

Effects of PBTs may include increase in disease, lesions in benthic feeders, spawning loss, change in age-structured populations of fish, and tissue contamination in fish and shellfish. Humans and other organisms, which consume shellfish and fish contaminated with persistent bioaccumulative pollutants, have the potential to bioaccumulate these chemicals. This may put these organisms at risk of mutagenic, teratogenic, and carcinogenic effects. Correlations have been found between elevated exposure to PCB mixtures and alterations in liver enzymes, hepatomegaly, and dermatological effects such as rashes have been reported.

Mercury

Inorganic

Inorganic mercury (elemental mercury) is less bioavailable and less toxic than that of organic mercury but is still toxic nonetheless. It is released into the environment through both natural sources as well as human activity, and it has the capability to travel long distances through the atmosphere. Around 2,700 to 6,000 tons of elemental mercury are released via natural activity such as volcanoes and erosion. Another 2,000 – 3,000 tons are released by human industrial activities such as coal combustion, metal smelting and cement production. Due to its high volatility and atmospheric residence time of around 1 year, mercury has the ability to travel across continents before being deposited. Inorganic mercury has a wide spectrum of toxicological effects that include damage to the respiratory, nervous, immune and excretory systems in humans. Inorganic mercury also possesses the ability to bioaccumulate individuals and biomagnify through trophic systems.

Organic

Organic mercury is significantly more detrimental to the environment than its inorganic form due to its widespread distribution as well as its higher mobility, general toxicity and rates of bioaccumulation than that of the inorganic form. Environmental organic mercury is mainly created by the transformation of elemental (inorganic) mercury via anaerobic bacteria into methylated mercury (organic). The global distribution of organic mercury is the result of general mobility of the compound, activation via bacteria and transportation from animal consumption. Organic mercury shares a lot of the same effects as the inorganic form but it has a higher toxicity due to its higher mobility in the body, especially its ability to readily move across the blood brain barrier.

Ecological Impact of Hg

The high toxicity of both forms of mercury (especially organic mercury) poses a threat to almost all organisms that comes in contact with it. This is one of the reasons that there is such high attention to mercury in the environment but even more so than its toxicity is both its persistence and atmospheric retention times. The ability of mercury to readily volatilize allows it to enter the atmosphere and travel far and wide. Unlike most other PBTs that have atmospheric half-lives between 30 min and 7 days mercury has an atmospheric residence time of at least 1 year. This atmospheric retention time along with mercury's resistance to degradation factors such as electromagnetic radiation and oxidation, which are two of the main factors leading to degradation of many PBTs in the atmosphere, allows mercury from any source to be transported extensively. This characteristic of mercury transportation globally along with its high toxicity is the reasoning behind its incorporation into the BNS list of PBTs.

Notable PBT Environmental Impacts

The realization of the adverse effects from environmental pollution were made public from several disasters that occurred globally. In 1965, it was recognized that extensive mercury pollution by the Chisso chemical factory in Minamata, Japan due to improper handling of industrial wastes resulted in significant effects to the humans and organisms exposed. Mercury was released into the environment as methyl mercury (bioavailable state) into industrial wastewater and was then bioaccumulated by shellfish and fish in Minamata Bay and the Shiranui Sea. When the contaminated seafood was consumed by the local populace it caused a neurological syndrome, coined Minamata disease. Symptoms include general muscle weakness, hearing damage, reduced field of vision, and ataxia. The Minamata disaster contributed to the global realization of the potential dangers from environmental pollution and to the characterization of PBTs.

Puget Sound

Despite the ban on DDT 30 years earlier and years of various efforts to clean up Puget Sound from DDT and PCB's, there is still a significant presence of both compounds which pose a constant threat to human health and the environment. Harbor seals (*Phoca vitulina*), a common marine species in the Puget Sound area have been the focus of a few studies to monitor and examine the effects of DDT accumulation and magnification in aquatic wildlife. One study tagged and reexamined seal pups every 4 to 5 years to be tested for DDT concentrations. The trends showed the pups to be highly contaminated; this means their prey are also highly contaminated. Due to DDT's high lipid solubility, it also has the ability to accumulate in the local populace who consume seafood from the area. This also translates to women who are pregnant or breastfeeding, since DDT will be transferred from the mother to child. Both animal and human health risk to DDT will continue to be an issue in Puget Sound especially because of the cultural significance of fish in this region.

Microtox Bioassay

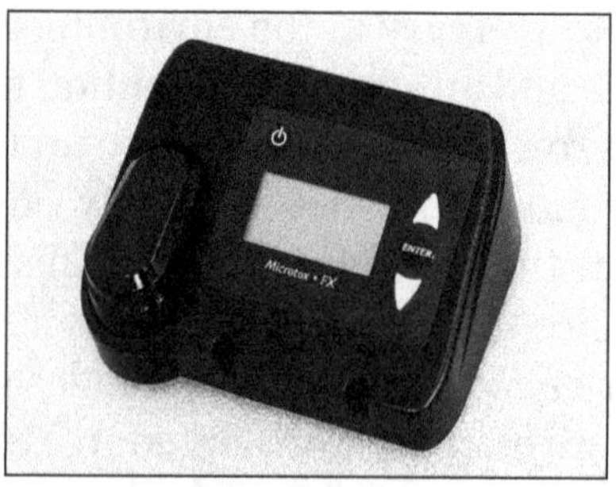

Microtox FX.

Microtox is an in vitro testing system which uses bioluminescent bacteria (*Allivibrio fischeri, formerly known as Vibrio fischeri*) to detect toxic substances in different substrates such as water, air, soils and sediments. *Allivibrio fischeri* are non-pathogenic, marine, bacteria that luminesce as a natural part of their metabolism. When exposed to a toxic substance, the respiratory process of the bacteria is disrupted, reducing light output. *Allivibrio fischeri* have demonstrated high sensitivity across a wide variety of toxic substances. Response to toxicity is observed as a change in luminescence, which is a by-product of cellular respiration. This change can be used to calculate a percent inhibition of *Allivibrio fischeri* that directly correlates to toxicity.

Microtox was developed by Azur Environmental (formerly Microbics Corporation) in 1979 as a cost effective alternative to toxicity tests available at the time. Prior to Microtox, the majority of toxicity tests available for water specifically focused on fish and daphnids. Since its inception, Microtox has become a standard method for testing the toxicity of water as well as other substrates such as soils and sediments.

In recent years the Microtox technology and name have undergone various different ownerships. In 2011, Microtox and related technologies was acquired by Modern Water from Strategic Diagnostics Incorporated (SDIX) for approximately $4.5 million. Prior to SDIX, Microtox was owned by its original developers Azur Environmental.

Microtox utilizes a bioluminescent bacteria (*Allivibrio fischeri*) to determine the toxicity of a particular substance and substrate. During cellular metabolism, these bacteria naturally emit light as a part of cellular respiration, which can be measured as luminescence. When exposed to toxic substances, a decrease in luminescence can be observed and percent change in luminescence can be directly correlated to toxicity. *Allivibrio fischeri* were specifically chosen, as these bacteria could be preserved by freeze-drying to increase shelf life and use. Both solid phase (soils and sediment) and aqueous acute toxicity testing can be conducted using this technology.

Materials

Analysers

The Microtox Model 500 is a laboratory-based photometer that measures acute toxicity. This analyser is a temperature-controlled, self-calibrating biosensor measuring system that uses the bioluminescence of *Allivibrio fischeri* to determine the toxicity of contaminated water, or elutriates of contaminated soils and sediments.

Microtox Continuous Toxicity Monitor (CTM) is a site-specific Microtox analyser that continuously measures the toxicity of a water source and provides results instantly. This fully automatic analyser has a broad detection range that can identify several thousand contaminants simultaneously whether or not there is knowledge of the source of contamination. This device has the ability to run continuously for up to 4 weeks, and is easy to operate and maintain.

The DeltaTox II is a portable instrument that can be used to conduct acute toxicity and adenosine triphosphate (ATP) testing. Also known as the portable version of the Microtox Model 500, this device provides simple testing, uses small sample sizes, and is a cost effective approach to analyzing water samples. This sensitive and rapid testing analyser has the ability to detect microbial contamination, as well as more than 2,700 different chemicals.

Reagent and Solutions

Microtox Model 500 and Microtox FX

The shelf life for the Acute Reagent is two years and for the solutions is three years when stored properly.

Microtox Acute Reagent is a freeze-dried culture of *Allivibrio fischeri* that is reconstituted prior to testing. It is recommended that the reagent be used within three hours of reconstitution. The sensitivity of the reagent may become altered after the recommended time period.

Microtox Osmotic Adjustment Solution (MOAS) is a nontoxic solution that is made up of 22% Sodium Chloride (NaCl) and Ultra Pure Water. This solution is added to a sample to adjust the osmotic pressure to approximately 2% NaCl.

Reconstitution Solution consists of specially prepared, nontoxic Ultra Pure Water.

Diluent is a nontoxic solution that is made up of 2% NaCl in Ultra Pure Water. This solution is used for diluting the sample and the reagent, and also provides osmotic protection that is required by *Allivibrio fischeri*.

Methods

Preparation of Samples

Microtox can be applied to a variety of matrices including drinking water, stormwater runoff, effluent, industrial discharges, soils and sediments. Most samples do not require special preparation before testing besides adjusting the salinity to 2%. However, samples that have certain characteristics, such as high turbidity levels, may require special preparation. If samples require a salinity adjustment to lower the salinity, this can be accomplished by adding an appropriate amount of Microtox Osmotic Adjusting Solution to dilute the sample. For example, adding 0.1 mL of MOAS to 1 mL of sample would result in a dilution of 90.9% of the original concentration. If a greater salinity is required, this can be accomplished by dissolving solid sodium chloride in the sample to achieve a final salinity of 2% for the protection of *Allivibrio fischeri*. Highly turbid samples that contain particulate matter will be required to settle before the test can be conducted. Particulate matter in the sample can interfere with bioluminescence by absorbing light and give misleading test results. Interference of luminescence can also occur with samples which are highly colored (particularly red, brown or black). It may be necessary to centrifuge samples to obtain an acceptable clarity for the test. If samples contain chlorine, this may alter the toxicity to *Allivibrio fischeri* and also give misleading results. The samples can be de-chlorinated using a sodium thiosulphate and deionized water solution that does not affect test results. Ideally, the pH of samples should not be modified since it is preferable to test each sample at the original pH level. However, if it is necessary to adjust the pH this should be done by adding either sodium hydroxide solution or hydrochloric acid to the sample.

Unlike water samples, soil and sediment samples are not homogenous. As a result, it is difficult to obtain representative samples from such matrices. Toxic substances are likely to bind to particulate matter, and the extent to which toxic materials bind depends on the composition of the particles. For example, smaller particles such as clay tend to tightly bind to chemicals, acting like ion exchange resins. Microtox tests for sediment and soil differ in the way the matrix is prepared for contact with *Allivibrio fischeri.* To obtain a representative soil or sediment sample, it is necessary to conduct an elutriate test. Sediment elutriates can be prepared through extraction with either distilled water, saline water, or an organic solvent such as methylene chloride, or hexane. To run an elutriate test a soil sample is mixed with an extractant for a period of time, then allowed to settle and a sample is taken from the extract. If particulate matter is in the sample that was collected, it may be necessary to centrifuge the sample for optimal clarity. Additionally, the pore water of sediments can be collected by centrifugation and tested without extraction.

Procedures

There are five major Microtox tests including the Basic Test, the 100% Test, the Solid Phase Test, the Comparison Test, and the Inhibition Test. Of these five tests, three are used for sediment and soil testing including the Basic Test, the 100% Test and the Solid-Phase Test. All of these versions follow the same general method of reconstituting the *Allivibrio fischeri* reagent in the Reconstitution Solution. Corrections are made for salinity and particulate matter, then the bacteria are exposed to the sample solution depending on the methods of the particular test. The light output of the bacteria is measured using a photometer after five and 15 minutes from exposing the bacteria to the samples. The light measured directly correlates to the toxicity of the sample, producing data that allows for the calculation of EC50 or IC50s, or other ECxx and ICxx values.

Acute Toxicity Basic Test is a procedure that measures the relative acute toxicity of a sample. This test is the best protocol for testing samples of unknown toxicity, a high level of toxicity, or when the test results are required to provide the highest confidence and precision. This test consists of two controls and eight sample dilutions in duplicate.

Acute Toxicity 100% Test is a procedure that tests the sample at 100% sample concentration and as a result includes adding reagent solution directly to the sample. This test is used for samples that are expected to have a low level of toxicity and is generally used as an environmental screening tool. Compared to the Basic Test it is more sensitive to operator technique, and as a result may be less precise.

Acute Toxicity Solid-Phase Test is a procedure that allows the test organism to come in direct contact with the solid sample as particulate in an aqueous suspension. Normally, this test provides results indicating equal or higher toxicity when compared to eluate or pore water tests of the same sample. This is due to either equal or increased bioavailability resulting from direct contact. This test is subject to several sources of interference of luminescence including loss of bacteria from effects other than toxicity such as filtration of the sample; absorption of light due to color; and scattering of light due to turbidity. Corrections can be made by testing a sample of similar particle composition that is known to not be toxic. This test consists of two controls and 13 sample dilutions in duplicate. The Solid-Phase Test exposes the bacteria in such a way that is not always possible with pore water and elutriate.

Acute Toxicity Comparison & Inhibition Tests are the best procedures for testing samples with a low level of toxicity when an ECxx can not be determined using the Basic Test. These protocols are recommended for testing waste water treatment plant effluent, stormwater runoff, drinking water, pore water, and eluate. These tests use multiple replicates of a sample at a single concentration. Similar to the Basic Test, the Comparison Test protocol uses zero time light readings used for correcting the timed light level readings. The Inhibition Test procedure does not use zero time light readings and therefore, can not use a correction factor for the timed light level readings. Both of these tests consists of five controls and five replicates of the sample at a single concentration.

Microtox Omni Software

The Microtox Omni Software was developed by Azur Environmental and allows users of the Microtox Model 500 Analyser to run tests, visualize data, calculate statistics and generate reports. This program contains a set of templates for all of the commonly used toxicity tests and allows you to modify or add to the provided templates. The modification of these templates allows for the production of new test formats not originally included with the software. The test templates that are included in this software define all of the parameters for a particular test method. This program calculates the most efficient way of setting up the desired test on the Model 500 Analyser. A test tutor is also included with Microsoft Omni that gives listed instructions on how to set up and run the test of interest. This software allows users to load files from previous versions of the Microtox DOS Software and also gives users the ability to save new data in that original format. A database is included in this software that grants users to access data from a number of other users and test sites, allowing for a comparison of data and trend changes over time. Microtox Omni can be used with any number of databases.

Application

Microtox has a variety of environmental and industrial applications. Common applications are for testing the toxicity of both marine and freshwater, as well as sediments for pesticides and other inorganic and organic chemicals.

- Drinking Water: Microtox is used to test drinking water sources in many areas where either accidental or deliberate contamination is possible. Toxic contaminants in drinking water are indicated by a change in the color or intensity of light, or by a change in the rate of oxygen use.

- Lakes and Rivers: Microtox is used to test the toxicity of lake and river sediments contaminated by metals or nonspecific contaminants. The Solid-Phase test is used for sediments, while either the Basic test or the 100% test is used for pore water.
- Sediment testing: Microtox is used to test and evaluate the toxicity of various marine and freshwater sediments contaminated by metals and organics. Aqueous extracts of contaminated soils, drilling muds, and sludge. Microtox data may be used to establish apparent effects threshold (AET), sediment quality standards and used for NPDES permits, as well as Superfund cleanup levels.
- Industrial: This bioassay is used in the assessment of the toxicity of complex industrial effluent sources. It is a cost effective way for monitoring and testing large numbers of samples. Microtox can also be applied as an early warning system (EWS) and aid in detecting the presence of toxic materials, as well as predict the outcomes of other bioassays and tests.

Microtox has also been applied to animal testing as an in vitro alternative.

Ecological Relevance

Numerous studies and published data comparing Microtox results with toxicity values for fish, crustaceans and algae have found a positive correlation. However, others have pointed out that the effect of luminosity on the survival of organisms is unknown. Concerns have also been expressed regarding the use of sediment extracts and not the sediment itself. It is possible that only water-soluble contaminants will be tested for, and therefore may not be representative of the full range of contaminants present in the sediment. Extracts may also remove contaminants that are not bioavailable. This could lead to an over or under estimation of contaminants and their biological effects.

Dead Zone

Dead zones are low-oxygen, or hypoxic, areas in the world's oceans and lakes. Because most organisms need oxygen to live, few organisms can survive in hypoxic conditions. That is why these areas are called dead zones.

Dead zones occur because of a process called eutrophication, which happens when a body of water gets too many nutrients, such as phosphorus and nitrogen. At normal levels, these nutrients feed the growth of an organism called cyanobacteria, or blue-green algae. With too many nutrients, however, cyanobacteria grows out of control, which can be harmful. Human activities are the main cause of these excess nutrients being washed into the ocean. For this reason, dead zones are often located near inhabited coastlines.

Understanding the eutrophication process provides the clearest picture of how and why dead zones develop.

Causes of Eutrophication

Eutrophic events have increased because of the rapid rise in intensive agricultural practices, industrial activities, and population growth. These three processes emit large amounts of nitrogen and phosphorous. These nutrients enter our air, soil, and water. Human activities have emitted nearly twice as much nitrogen and three times as much phosphorus as natural emissions.

Different regions of the world emit different levels of these nutrients. In developed countries, such as the United States and nations in the European Union, heavy use of animal manure and commercial fertilizers in agricultureare the main contributors to eutrophication. Runoff from large agricultural fields enters creeks and bays because of rain or irrigation practices.

In developing countries of Latin America, Asia, and Africa, untreated wastewater from sewage and industry mainly contribute to eutrophication. Factories and sewage facilities are less regulated than they are in developed countries, and sometimes wastewater is simply dumped into creeks, rivers, lakes, or the ocean.

Atmospheric sources of nitrogen also contribute to eutrophication in some areas of the world. Fossil fuels and fertilizers release nitrogen into the atmosphere. This atmospheric nitrogen is then redeposited on land and water through the water cycle—rain and snow.

The Chesapeake Bay, on the East Coast of the United States, has one of the first dead zones ever identified, in the 1970s. The Chesapeake's high levels of nitrogen are caused by two factors: urbanization and agriculture. The western part of the bay is full of factories and urban centers that emit nitrogen into the air. Atmospheric nitrogen accounts for about a third of the nitrogen that enters the bay. The eastern part of the bay is a center of poultry farming, which produces large amounts of manure.

Since 1967, the Chesapeake Bay Foundation has led a number of programs that aim to improve the bay's water quality and curb pollution runoff. The Chesapeake still has a dead zone, whose size varies with the season and weather.

Eutrophication and the Environment

The eutrophication process has severe environmental impacts. Dead zones result from these impacts, which include algal blooms and hypoxia.

Algal Blooms

Phosphorous, nitrogen, and other nutrients increase the productivity or fertility of marine ecosystems. Organisms such as phytoplankton, algae, and seaweeds will grow quickly and excessively on the water's surface. This rapid development of algae and phytoplankton is called an algal bloom. Algal blooms can create dead zones beneath them.

Algal blooms prevent light from penetrating the water's surface. They also prevent oxygen from being absorbed by organisms beneath them. Sunlight is necessary for plants and organisms like phytoplankton and algae, which manufacture their own nutrients from sunlight, water, and carbon dioxide. Oxygen is necessary for almost all aquatic life, from sea grasses to fish.

By depriving organisms of sunlight and oxygen, algal blooms negatively impact a variety of species that live below the water's surface. The number and diversity of benthic, or bottom-dwelling, species are especially reduced.

Because algae dominates the aquatic ecosystem, algal blooms are sometimes referred to as "red tides" or "brown tides," depending on the color of the algae. Red tides actually have nothing to do with tides. They also have nothing to do with algae. The organism that causes red tides is a bacteria, called cyanobacteria.

Algal blooms also cause larger-scale problems, such as human illness. Shellfish, such as oysters, are filter feeders. As they filter water, they absorb microbes associated with algal blooms. Many of these microbes are toxic to people. People may become sick or even die from shellfish poisoning.

Algal blooms can also lead to the death of marine mammals and shore birds that rely on the marine ecosystem for food. Wading birds, such as herons, and mammals, such as sea lions, depend on fish for survival. With fewer fish beneath algal blooms, these animals lose an important food source.

Algal blooms can also impact aquaculture, or the farming of marine life. One red tide event wiped out 90 percent of the entire stock of Hong Kong's fish farms in 1998, resulting in an estimated economic loss of $40 million.

Algal blooms usually die soon after they appear. The ecosystem simply cannot support the huge number of cyanobacteria. The organisms compete with one another for the remaining oxygen and nutrients.

Hypoxia

Hypoxia occurs when algae and other organisms die from lack of oxygen and available nutrients. Hypoxia events often follow algal blooms. The cyanobacteria, algae, and phytoplankton sink to the seafloor, and are decomposed by bacteria. Even though oxygen can now flow freely through the aquatic ecosystem, the decomposition process uses up almost all of it. This lack of oxygen creates dead zones in which most aquatic species cannot survive.

The Gulf of Mexico has a seasonal hypoxic zone that forms every year in late summer. Its size varies from fewer than 5,000 square kilometers (1,931 square miles) to approximately 22,000 square kilometers (8,494 square miles, or the size of Massachusetts). Concern over its increasing size led to the formation of the Mississippi River/Gulf of Mexico Watershed Nutrient Task Force in 1997. Its mission is to reduce the five-year running average of the Gulf of Mexico dead zone to less than 5,000 square kilometers.

The Baltic Sea is home to seven of the world's 10 largest marine dead zones. Increased runoff from agricultural fertilizers and sewage has sped up the eutrophication process. Overfishing of Baltic cod has intensified the problem. Cod eat sprats, a small, herring-like species that eat microscopic zooplankton, which in turn eat algae. Fewer cod and more sprats mean more algae and less oxygen. The spreading dead zones are starting to reach the cod's deep-water breeding grounds, further endangering the species.

The Baltic Sea has become the first "macro-region" targeted by the European Union to combat pollution, dead zones, overfishing, and regional disputes. The EU is coordinating the Baltic Sea Strategy with eight EU member countries that border the Baltic Sea: Denmark, Estonia, Finland, Germany, Latvia, Lithuania, Poland, and Sweden.

Categorizing Eutrophic Systems

Scientists have identified 415 dead zones worldwide. Hypoxic areas have increased dramatically during the past 50 years, from about 10 documented cases in 1960 to at least 169 in 2007. The majority of the world's dead zones are located along the eastern coast of the United States, and the coastlines of the Baltic States, Japan, and the Korean Peninsula.

As a result of the dramatic increase in dead zones, scientists have categorized coastal systems experiencing any symptoms of eutrophication.

The dead zone in the Gulf of Mexico is most dramatic near the delta of the Mississippi River.

An *area of concern* is a coastal system that exhibits effects of eutrophication, such as elevated nutrient levels, harmful algal blooms, and negative changes in the benthic community. Areas of concern are at the most risk of developing hypoxia. Major concentrations of areas of concern are located along the western coast of Central and South America, and the coastlines of Great Britain and Australia. There are 233 areas of concern around the world.

A *system in recovery* is one that once exhibited low oxygen levels and hypoxia, but is now improving. For example, the Black Sea once experienced annual hypoxic events, but is now in a state of recovery. Others, like Boston Harbor in the United States and the Mersey Estuary in the United Kingdom, also have improved water quality. These are the results of better industrial and wastewater controls. There are only 13 coastal systems in recovery around the world.

References

- Ecosystem-toxicology, topics-medicine-and-dentistry: sciencedirect.com, Retrieved 10 May, 2020
- Ecological, profiles-extoxnet-TIB: pmep.cce.cornell.edu, Retrieved 18 June, 2020
- El-Shahawi, M.S., Hamza, A., Bashammakhb, A.S., Al-Saggaf, W.T. (2010). An overview on the accumulation, distribution, transformations, toxicity and analytical methods for the monitoring of persistent organic pollutants. Talanta. 80, 1587–1597. doi:10.1016/j.talanta.2009.09.055
- Dead-zone, encyclopedia: nationalgeographic.org, Retrieved 17 July, 2020
- Richter, E.; Wick, A.; Ternes, T.A.; Coors, A. (2013). "Ecotoxicity of Climbzole, a fungicide contained in Antridandruff shampoo". Environmental Toxicology and Chemistry. 32 (12): 2816–2825. doi:10.1002/etc.2367
- Simon T, Britt JK, James RC (2007). "Development of a neurotoxic equivalence scheme of relative potency for assessing the risk of PCB mixtures". Regulatory Toxicology and Pharmacology 48(2): 148–70. DOI:10.1016/j.yrtph.2007.03.005. PMID 17475378

4
Aquatic Toxicology

Aquatic toxicology is associated with the study of the harmful effects of chemicals on aquatic organisms and their ecosystems. Its major factors include ocean acidification, oil spilling in water bodies, aquatic contaminants and pollutants, etc. The topics elaborated in this chapter will help in gaining a better perspective of aquatic toxicology.

Aquatic toxicology is the study of the effects of manufactured chemicals and other anthropogenic and natural materials and activities on aquatic organisms at various levels of organization, from subcellular through individual organisms to communities and ecosystems.

Aquatic toxicology generally involves the measurement of contaminant levels to characterize the hazards imposed on the aquatic environment; however, this field of study also includes information on how those contaminants can affect humans in and around these aquatic environments. The multidisciplinary research that comprises the field of aquatic toxicology has provided a better understanding of the effects anthropogenic activities and chemical contaminants have on aquatic environments. Additionally, this increase in knowledge has improved the methods utilized and consequently, the confidence in measuring the potential hazards associated with humans exposed to contaminated aquatic environments and organisms. The resources assembled are informative whether one is looking at the pollution generated by anthropogenic activities and the affected aquatic ecosystems or observing how contaminants in water and aquatic organisms affect water quality and the subsequent exposure to humans.

In aquatic toxicology, it is now well-established that the chemical characterization of pollutants exposure is not sufficient and that multidisciplinary approaches coupling chemistry and biology have to be developed to allow linking of the presence of contaminants and their putative toxic impacts. Along with the improvement of molecular biology methods, the development of "omics" technologies is booming worldwide since the early 1990s when these techniques have begun to emerge. During the past 20 years, these techniques have been in turn described as a unique research opportunity to decrypt all the biological mechanisms.

The goal of aquatic toxicology is to predict the effects of contaminants in ecosystems. This should form the basis of policies affecting aquatic systems. Predictions require that existing observations can be used to generate scenarios about the future. This is the case, although aquatic toxicological studies in the natural environment necessarily report what has already happened (thus being retrospective). This holds also for biomonitoring studies, which rely on it being possible to extrapolate from any observed trend to the future. The utility of retrospective studies is twofold. First, they can indicate how the environment has already been affected. Such information can be required in order to identify the sources of contamination, to require those responsible either to carry out stricter cleaning of their effluent or to pay for the damages incurred. Both necessitate the damaging contamination being pinpointed to a definite source, which requires highly specific exposure biomarkers. Second, they can be used to prevent similar discharges and effects in other places. This requires the assumption that other environmental factors do not significantly affect the contaminant responses that are observed. Laboratory studies, on the other hand, can use novel contaminants and their mixtures. Thus, possible contaminants can be studied before their appearance in the environment. However, these studies do not include all the contaminants, their interactions, and interactions with natural abiotic and biotic factors. Consequently, one must always assume that the factors studied in the laboratory are decisive for the effects in nature.

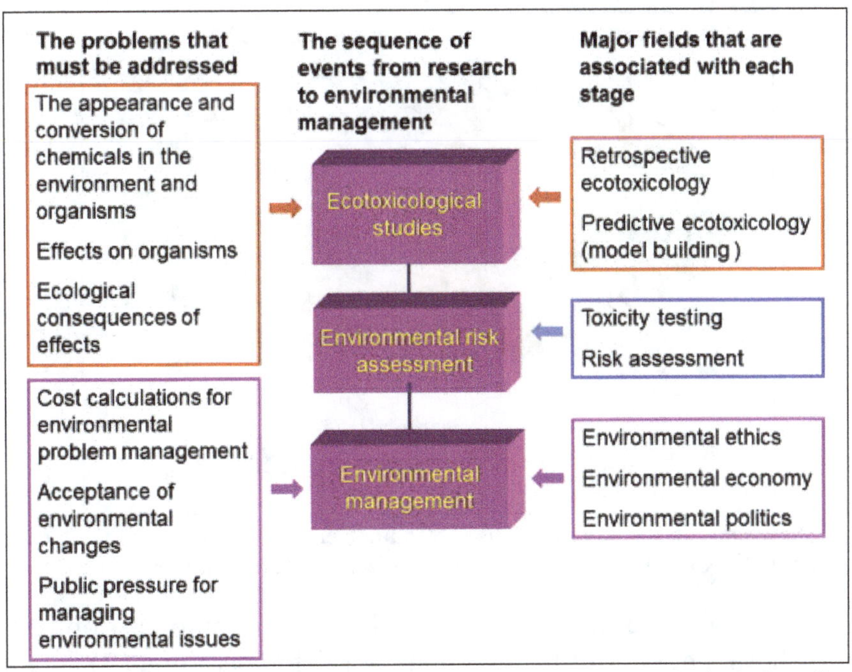

A scheme of how environmental management follows from ecotoxicological studies.

The same color in arrows and box outlines indicates the same stage in the decision-making pathway. Note that toxicity testing and risk assessment are not included in ecotoxicological research. For toxicity testing, this is because testing uses standard methods with the aim of making comparisons. Development of new testing methods is part of ecotoxicological research. Risk assessment itself is not research, but uses the data gathered in ecotoxicological studies to evaluate the risks that the contaminants assessed cause to the environment.

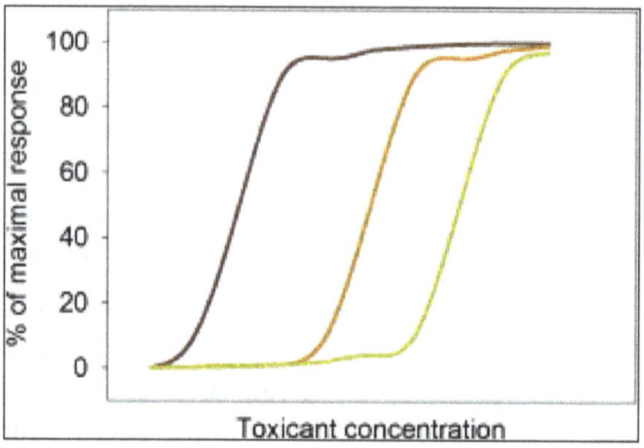

A toxicant may have several modes of action (targets), varying with concentration; different concentration–response profiles are depicted with lines of different colors.

Ocean Acidification

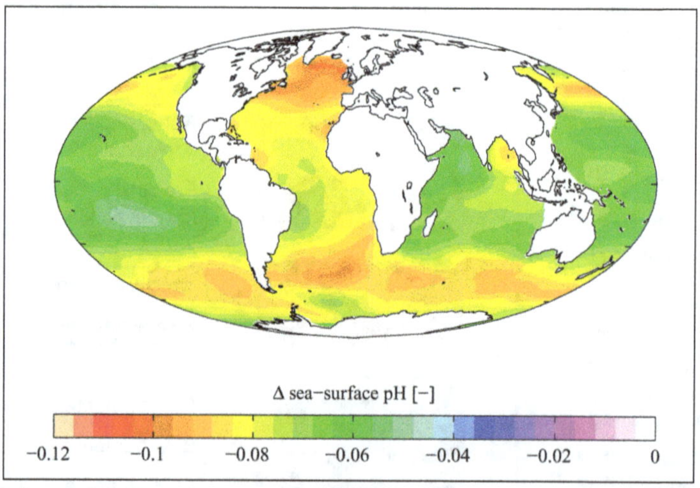

Estimated change in sea water pH caused by human created CO_2 between the 1700s and the 1990s, from the Global Ocean Data Analysis Project (GLODAP) and the World Ocean Atlas.

NOAA provides evidence for upwelling of "acidified" water onto the Continental Shelf. In the figure above, note the vertical sections of (A) temperature, (B) aragonite saturation, (C) pH, (D) DIC, and (E) pCO_2 on transect line 5 off Pt. St. George, California. The potential density surfaces are superimposed on the temperature section. The 26.2 potential density surface delineates the location of the first instance in which the undersaturated water is upwelled from depths of 150 to 200 m onto the shelf and outcropping at the surface near the coast. The red dots represent sample locations.

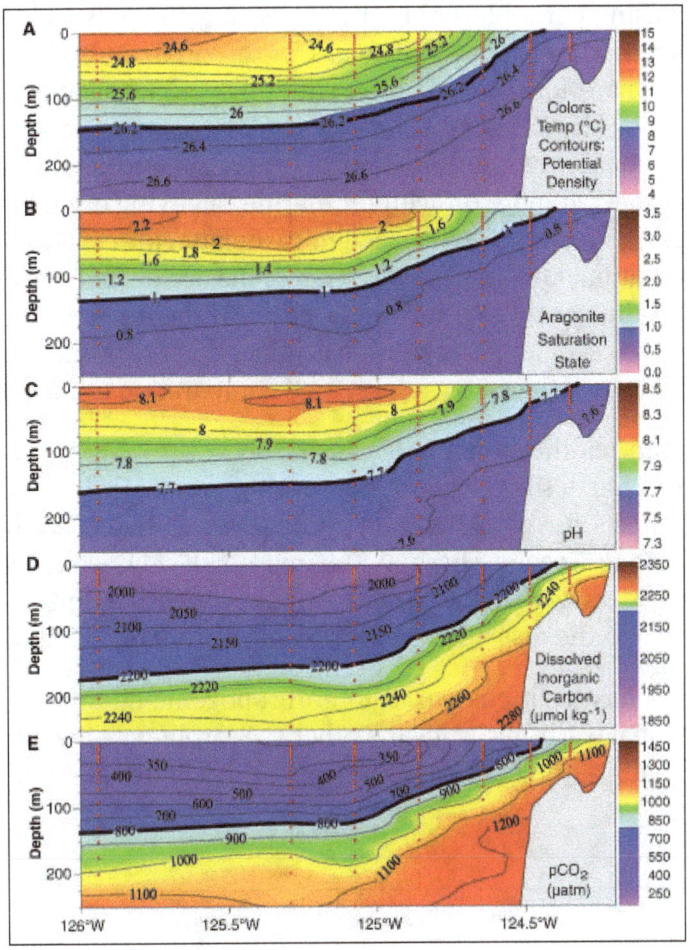

Ocean acidification is the ongoing decrease in the pH of the Earth's oceans, caused by the uptake of carbon dioxide (CO_2) from the atmosphere. Seawater is slightly basic (meaning pH > 7), and ocean acidification involves a shift towards pH-neutral conditions rather than a transition to acidic conditions (pH < 7). An estimated 30–40% of the carbon dioxide from human activity released into the atmosphere dissolves into oceans, rivers and lakes. Some of it reacts with the water to form carbonic acid. Some of the resulting carbonic acid molecules dissociate into a bicarbonate ion and a hydrogen ion, thus increasing ocean acidity (H^+ ion concentration). Between 1751 and 1996, surface ocean pH is estimated to have decreased from approximately 8.25 to 8.14, representing an increase of almost 30% in H^+ ion concentration in the world's oceans. Earth System Models project that, by around 2008, ocean acidity exceeded historical analogues and, in combination with other ocean biogeochemical changes, could undermine the functioning of marine ecosystems and disrupt the provision of many goods and services associated with the ocean beginning as early as 2100.

Increasing acidity is thought to have a range of potentially harmful consequences for marine organisms such as depressing metabolic rates and immune responses in some organisms and causing coral bleaching. By increasing the presence of free hydrogen ions, the additional carbonic acid that forms in the oceans ultimately results in the conversion of carbonate ions into bicarbonate ions. Ocean alkalinity (roughly equal to $[HCO_3^-] + 2[CO_3^{2-}]$) is not changed by the process, or may increase over long time periods due to carbonate dissolution. This net decrease in the amount of carbonate ions available may make it more difficult for marine calcifying organisms, such as coral and some plankton, to form biogenic calcium carbonate, and such structures become vulnerable to dissolution. Ongoing acidification of the oceans may threaten future food chains linked with the oceans. As members of the InterAcademy Panel, 105 science academies have issued a statement on ocean acidification recommending that by 2050, global CO_2 emissions be reduced by at least 50% compared to the 1990 level.

While ongoing ocean acidification is at least partially anthropogenic in origin, it has occurred previously in Earth's history, and the resulting ecological collapse in the oceans had long-lasting effects for global carbon cycling and climate. The most notable example is the Paleocene-Eocene Thermal Maximum (PETM), which occurred approximately 56 million years ago when massive amounts of carbon entered the ocean and atmosphere, and led to the dissolution of carbonate sediments in all ocean basins.

Ocean acidification has been compared to anthropogenic climate change and called the "evil twin of global warming" and "the other CO_2 problem". Freshwater bodies also appear to be acidifying, although this is a more complex and less obvious phenomenon.

Carbon Cycle

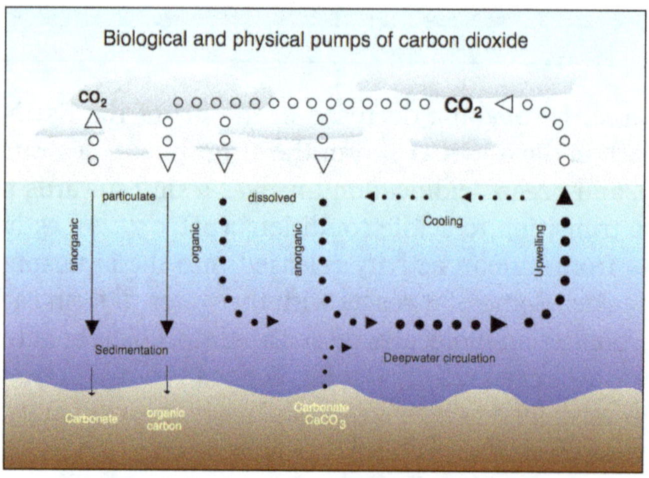

The CO_2 cycle between the atmosphere and the ocean.

The carbon cycle describes the fluxes of carbon dioxide (CO_2) between the oceans, terrestrial biosphere, lithosphere, and the atmosphere. Human activities such as the combustion of fossil fuels and land use changes have led to a new flux of CO_2 into the atmosphere. About 45% has remained in the atmosphere; most of the rest has been taken up by the oceans, with some taken up by terrestrial plants.

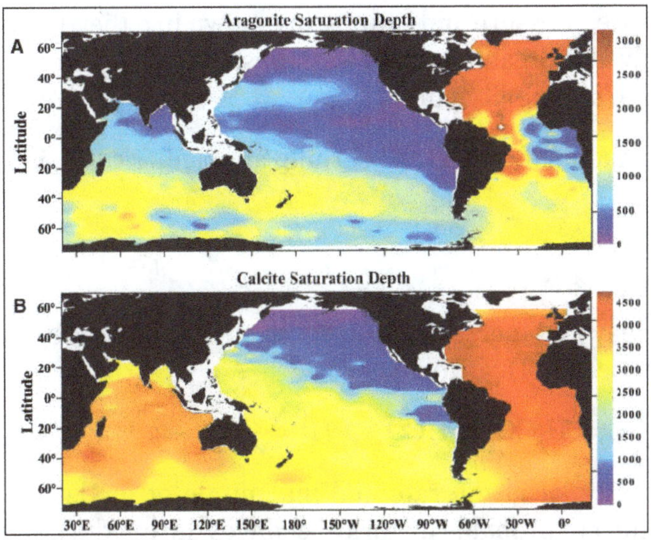

Distribution of (A) aragonite and (B) calcite saturation depth in the global oceans.

This map shows changes in the aragonite saturation level of ocean surface waters between the 1880s and the most recent decade. Aragonite is a form of calcium carbonate that many marine animals use to build their skeletons and shells. The lower the saturation level, the more difficult it is for organisms to build and maintain their skeletons and shells. A negative change represents a decrease in saturation.

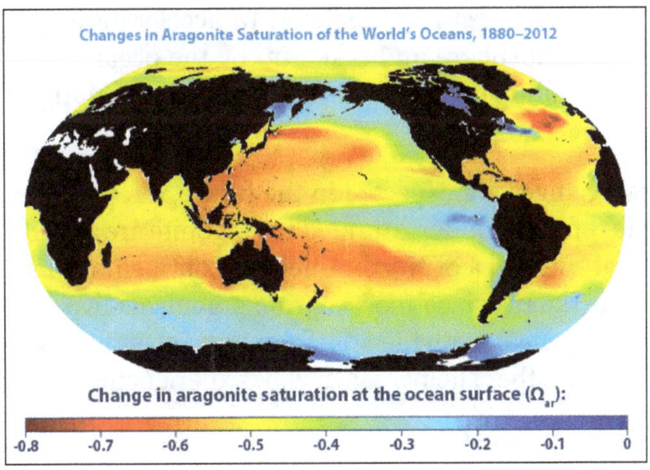

The carbon cycle involves both organic compounds such as cellulose and inorganic carbon compounds such as carbon dioxide, carbonate ion, and bicarbonate ion. The inorganic compounds are particularly relevant when discussing ocean acidification for they include many forms of dissolved CO_2 present in the Earth's oceans.

When CO_2 dissolves, it reacts with water to form a balance of ionic and non-ionic chemical species: dissolved free carbon dioxide ($CO_{2(aq)}$), carbonic acid (H_2CO_3), bicarbonate (HCO^-_3) and carbonate (CO^{2-}_3). The ratio of these species depends on factors such as seawater temperature, pressure and salinity (as shown in a Bjerrum plot). These different forms of dissolved inorganic carbon are transferred from an ocean's surface to its interior by the ocean's solubility pump.

The resistance of an area of ocean to absorbing atmospheric CO_2 is known as the Revelle factor.

Acidification

Dissolving CO_2 in seawater increases the hydrogen ion (H^+) concentration in the ocean, and thus decreases ocean pH, as follows:

$$CO_{2\,(aq)} + H_2O \rightleftharpoons H_2CO_3 \rightleftharpoons HCO_3^- + H^+ \rightleftharpoons CO_3^{2-} + 2\,H^+.$$

Caldeira and Wickett placed the rate and magnitude of modern ocean acidification changes in the context of probable historical changes during the last 300 million years.

Since the industrial revolution began, the ocean has absorbed about a third of the CO_2 we have produced since then and it is estimated that surface ocean pH has dropped by slightly more than 0.1 units on the logarithmic scale of pH, representing about a 29% increase in H^+. It is expected to drop by a further 0.3 to 0.5 pH units (an additional doubling to tripling of today's post-industrial acid concentrations) by 2100 as the oceans absorb more anthropogenic CO_2, the impacts being most severe for coral reefs and the Southern Ocean. These changes are predicted to accelerate as more anthropogenic CO_2 is released to the atmosphere and taken up by the oceans. The degree of change to ocean chemistry, including ocean pH, will depend on the mitigation and emissions pathways taken by society.

Although the largest changes are expected in the future, a report from NOAA scientists found large quantities of water undersaturated in aragonite are already upwelling close to the Pacific continental shelf area of North America. Continental shelves play an important role in marine ecosystems since most marine organisms live or are spawned there, and though the study only dealt with the area from Vancouver to Northern California, the Researchers suggest that other shelf areas may be experiencing similar effects.

Table: Average surface ocean pH.

Time	pH	pH change relative to pre-industrial	Source	H^+ concentration change relative to pre-industrial
Pre-industrial (18th century)	8.179		analysed field	
Recent past (1990s)	8.104	−0.075	field	+ 18.9%
Present levels	~8.069	−0.11	field	+ 28.8%
2050 (2×CO_2 = 560 ppm)	7.949	−0.230	model	+ 69.8%
2100 (IS92a)	7.824	−0.355	model	+ 126.5%

Rate

One of the first detailed datasets to examine how pH varied over 8 years at a specific north temperate coastal location found that acidification had strong links to *in situ* benthic species dynamics and that the variation in ocean pH may cause calcareous species to perform more poorly than noncalcareous species in years with low pH and predicts consequences for near-shore benthic ecosystems. Thomas Lovejoy, former chief biodiversity advisor to the World Bank, has suggested that "the acidity of the oceans will more than double in the next 40 years. He says this rate is 100 times faster than any changes in ocean acidity in the last 20 million years, making it unlikely that marine life can somehow adapt to the changes." It is predicted that, by the year 2100, If co-occurring biogeochemical changes influence the delivery of ocean goods and services, then they could also have a considerable effect on human welfare for those who rely heavily on the ocean for food, jobs, and revenues. A panel of experts who had previously participated in the IPCC reports have determined that it is not yet possible to determine a threshold for ocean acidity that should not be exceeded.

Current rates of ocean acidification have been compared with the greenhouse event at the Paleocene–Eocene boundary (about 55 million years ago) when surface ocean temperatures rose by 5–6 degrees Celsius. No catastrophe was seen in surface ecosystems, yet bottom-dwelling organisms in the deep ocean experienced a major extinction. The current acidification is on a path to reach levels higher than any seen in the last 65 million years, and the rate of increase is about ten times the rate that preceded the Paleocene–Eocene mass extinction. The current and projected acidification has been described as an almost unprecedented geological event. A National Research Council study released in April 2010 likewise concluded that "the level of acid in the oceans is increasing at an unprecedented rate". A 2012 paper in the journal *Science* examined the geological record in an attempt to find a historical analog for current global conditions as well as those of the future. The researchers determined that the current rate of ocean acidification is faster than at any time in the past 300 million years.

A review by climate scientists by the Royal Society of the UK similarly highlighted the centrality of the *rates* of change in the present anthropogenic acidification process, writing:

> "The natural pH of the ocean is determined by a need to balance the deposition and burial of $CaCO_3$ on the sea floor against the influx of Ca^{2+} and CO^{2-}_3 into the ocean from dissolving rocks on land, called weathering. These processes stabilize the pH of the ocean, by a mechanism called $CaCO_3$ compensation.The point of bringing it up again is to note that if the CO_2 concentration of the atmosphere changes more slowly than this, as it always has throughout the Vostok record, the pH of the ocean will be relatively unaffected because $CaCO_3$ compensation can keep up. The [present] fossil fuel acidification is much faster than natural changes, and so the acid spike will be more intense than the earth has seen in at least 800,000 years."

In the 15-year period 1995–2010 alone, acidity has increased 6 percent in the upper 100 meters of the Pacific Ocean from Hawaii to Alaska. According to a statement in July 2012 by Jane Lubchenco, head of the U.S. National Oceanic and Atmospheric Administration "surface waters are changing much more rapidly than initial calculations have suggested. It's yet another reason to be very seriously concerned about the amount of carbon dioxide that is in the atmosphere now and the additional amount we continue to put out."

A 2013 study claimed acidity was increasing at a rate 10 times faster than in any of the evolutionary crises in Earth's history. In a synthesis report published in *Science* in 2015, 22 leading marine scientists stated that CO_2 from burning fossil fuels is changing the oceans' chemistry more rapidly than at any time since the Great Dying, Earth's most severe known extinction event, emphasizing that the 2 °C maximum temperature increase agreed upon by governments reflects too small a cut in emissions to prevent "dramatic impacts" on the world's oceans, with lead author Jean-Pierre Gattuso remarking that "The ocean has been minimally considered at previous climate negotiations.

The rate at which ocean acidification will occur may be influenced by the rate of surface ocean warming, because the chemical equilibria that govern seawater pH are temperature-dependent. Greater seawater warming could lead to a smaller change in pH for a given increase in CO_2.

Calcification

Changes in ocean chemistry can have extensive direct and indirect effects on organisms and their habitats. One of the most important repercussions of increasing ocean acidity relates to the production of shells and plates out of calcium carbonate ($CaCO_3$). This process is called calcification and is important to the biology and survival of a wide range of marine organisms. Calcification involves the precipitation of dissolved ions into solid $CaCO_3$ structures, such as coccoliths. After they are formed, such structures are vulnerable to dissolution unless the surrounding seawater contains saturating concentrations of carbonate ions (CO_3^{2-}).

Mechanism

Of the extra carbon dioxide added into the oceans, some remains as dissolved carbon dioxide, while the rest contributes towards making additional bicarbonate (and additional carbonic acid). This also increases the concentration of hydrogen ions, and the percentage increase in hydrogen is larger than the percentage increase in bicarbonate, creating an imbalance in the reaction $HCO_3^- \rightleftharpoons CO_3^{2-} + H^+$. To maintain chemical equilibrium, some of the carbonate ions already in the ocean combine with some of the hydrogen ions to make further bicarbonate. Thus the ocean's concentration of carbonate ions is reduced, creating an imbalance in the reaction $Ca^{2+} + CO_3^{2-} \rightleftharpoons CaCO_3$, and making the dissolution of formed $CaCO_3$ structures more likely.

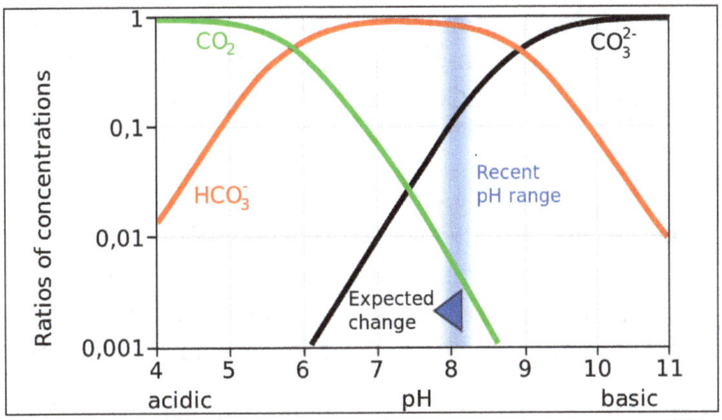

Bjerrum plot: Change in carbonate system of seawater from ocean acidification.

The increase in concentrations of dissolved carbon dioxide and bicarbonate, and reduction in carbonate, are shown in a Bjerrum plot.

Saturation State

The saturation state (known as Ω) of seawater for a mineral is a measure of the thermodynamic potential for the mineral to form or to dissolve, and for calcium carbonate is described by the following equation:

$$\Omega = \frac{\left[Ca^{2+}\right]\left[CO_3^{2-}\right]}{K_{sp}}.$$

Here Ω is the product of the concentrations (or activities) of the reacting ions that form the mineral (Ca^{2+} and CO^{2-}_3), divided by the product of the concentrations of those ions when the mineral is at equilibrium (K_{sp}), that is, when the mineral is neither forming nor dissolving. In seawater, a natural horizontal boundary is formed as a result of temperature, pressure, and depth, and is known as the saturation horizon. Above this saturation horizon, Ω has a value greater than 1, and $CaCO_3$ does not readily dissolve. Most calcifying organisms live in such waters. Below this depth, Ω has a value less than 1, and $CaCO_3$ will dissolve. However, if its production rate is high enough to offset dissolution, $CaCO_3$ can still occur where Ω is less than 1. The carbonate compensation depth occurs at the depth in the ocean where production is exceeded by dissolution.

The decrease in the concentration of CO_3^{2-} decreases Ω, and hence makes $CaCO_3$ dissolution more likely.

Calcium carbonate occurs in two common polymorphs (crystalline forms): aragonite and calcite. Aragonite is much more soluble than calcite, so the aragonite saturation horizon is always nearer to the surface than the calcite saturation horizon. This also means that those organisms that produce aragonite may be more vulnerable to changes in ocean acidity than those that produce calcite. Increasing CO_2 levels and the resulting lower pH of seawater decreases the saturation state of $CaCO_3$ and raises the saturation horizons of both forms closer to the surface. This decrease in saturation state is believed to be one of the main factors leading to decreased calcification in marine organisms, as the inorganic precipitation of $CaCO_3$ is directly proportional to its saturation state.

Possible Impacts

Increasing acidity has possibly harmful consequences, such as depressing metabolic rates in jumbo squid, depressing the immune responses of blue mussels, and coral bleaching. However it may benefit some species, for example increasing the growth rate of the sea star, *Pisaster ochraceus*, while shelled plankton species may flourish in altered oceans.

The reports "Ocean Acidification Summary for Policymakers 2013" and the IPCC approved "Special Report on the Ocean and Cryosphere in a Changing Climate" from 2019 describe research findings and possible impacts.

Impacts on Oceanic Calcifying Organisms

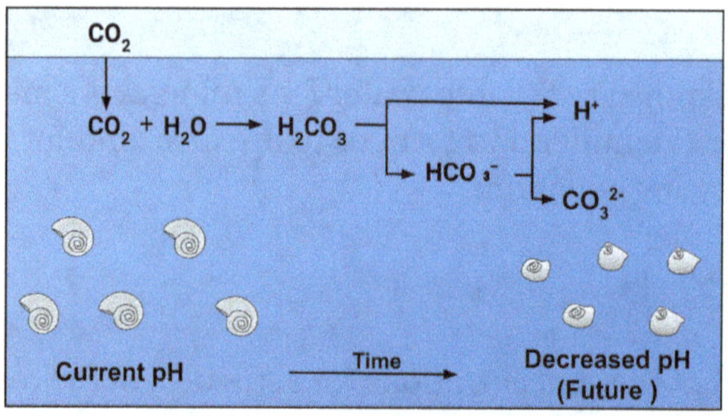

Shells of pteropods dissolve in increasingly acidic conditions caused by increased amounts of atmospheric CO_2.

Although the natural absorption of CO_2 by the world's oceans helps mitigate the climatic effects of anthropogenic emissions of CO_2, it is believed that the resulting decrease in pH will have negative consequences, primarily for oceanic calcifying organisms. These span the food chain from autotrophs to heterotrophs and include organisms such as coccolithophores, corals, foraminifera, echinoderms, crustaceans and molluscs. As described above, under normal conditions, calcite and aragonite are stable in surface waters since the carbonate ion is at supersaturating concentrations. However, as ocean pH falls, the concentration of carbonate ions required for saturation to occur increases, and when carbonate becomes undersaturated, structures made of calcium carbonate are vulnerable to dissolution. Therefore, even if there is no change in the rate of calcification, the rate of dissolution of calcareous material increases.

Corals, coccolithophore algae, coralline algae, foraminifera, shellfish and pteropods experience reduced calcification or enhanced dissolution when exposed to elevated CO_2.

The Royal Society published a comprehensive overview of ocean acidification, and its potential consequences, in June 2005. However, some studies have found different response to ocean acidification, with coccolithophore calcification and photosynthesis both increasing under elevated atmospheric pCO_2, an equal decline in primary production and calcification in response to elevated CO_2 or the direction of the response varying between species. A study in 2008 examining a sediment core from the North Atlantic found that while the species composition of coccolithophorids has remained unchanged for the industrial period 1780 to 2004, the calcification of coccoliths has increased by up to 40% during the same time. A 2010 study Suggested that while some areas are overharvested and other fishing grounds are being restored, because of ocean acidification it may be impossible to bring back many previous shellfish populations. While the full ecological consequences of these changes in calcification are still uncertain, it appears likely that many calcifying species will be adversely affected.

When exposed in experiments to pH reduced by 0.2 to 0.4, larvae of a temperate brittlestar, a relative of the common sea star, fewer than 0.1 percent survived more than eight days. There is also a suggestion that a decline in the coccolithophores may have secondary effects on climate, contributing to global warming by decreasing the Earth's albedo via their effects on oceanic cloud cover. All marine ecosystems on Earth will be exposed to changes in acidification and several other ocean biogeochemical changes.

The fluid in the internal compartments where corals grow their exoskeleton is also extremely important for calcification growth. When the saturation rate of aragonite in the external seawater is at ambient levels, the corals will grow their aragonite crystals rapidly in their internal compartments, hence their exoskeleton grows rapidly. If the level of aragonite in the external seawater is lower than the ambient level, the corals have to work harder to maintain the right balance in the internal compartment. When that happens, the process of growing the crystals slows down, and this slows down the rate of how much their exoskeleton is growing. Depending on how much aragonite is in the surrounding water, the corals may even stop growing because the levels of aragonite are too low to pump into the internal compartment. They could even dissolve faster than they can make the crystals to their skeleton, depending on the aragonite levels in the surrounding water. Under the current progression of carbon emissions, around 70% of North Atlantic cold-water corals will be living in corrosive waters by 2050–60.

A study conducted showed that the skeletal growth of corals under acidified conditions is primarily affected by a reduced capacity to build dense exoskeletons, rather than affecting the linear extension of the exoskeleton. Using Global Climate Models, they show that the density of some species of corals could be reduced by over 20% by the end of this century.

An *in situ* experiment on a 400 m^2 patch of the Great Barrier Reef to decrease seawater CO_2 level (raise pH) to close to the preindustrial value showed a 7% increase in net calcification. A similar experiment to raise *in situ* seawater seawater CO_2 level (lower pH) to a level expected soon after the middle of this century found that net calcification decreased 34%.

Ocean acidification may force some organisms to reallocate resources away from productive endpoints such as growth in order to maintain calcification.

In some places carbon dioxide bubbles out from the sea floor, locally changing the pH and other aspects of the chemistry of the seawater. Studies of these carbon dioxide seeps have documented a variety of responses by different organisms. Coral reef communities located near carbon dioxide seeps are of particular interest because of the sensitivity of some corals species to acidification. In Papua New Guinea, declining pH caused by carbon dioxide seeps is associated with declines in coral species diversity. However, in Palau carbon dioxide seeps are not associated with reduced species diversity of corals, although bioerosion of coral skeletons is much higher at low pH sites.

Ocean acidification may affect the ocean's biologically driven sequestration of carbon from the atmosphere to the ocean interior and seafloor sediment, weakening the so called biological pump. Seawater acidification could also see Antarctic phytoplanktons smaller and less effective at storing carbon.

Impact on Reef Fish

With the production of CO_2 from the burning of fossil fuels, oceans are becoming more acidic since CO_2 dissolves in water and forms the acidic bicarbonate ion. This results in a pH drop which then causes corals to expel their algae with which they have a symbiotic relationship with, causing the coral to eventually die due to a lack of nutrients.

Since corals reefs are one of the most diverse ecosystems on the planet, coral bleaching due to ocean acidification could result in a major loss of habitat for the many species of reef fish, resulting in increased predation and the eventual endangered classification or extinction of countless species. This will ultimately decrease the overall diversity of fish in marine environments, which will cause many predators of reef fish to die off since their normal supply of food was cut off. Food webs in coral reefs will also be greatly impacted because once a species goes extinct or is less prevalent, their natural predators will lose their primary food source causing the food web to collapse in on itself. If such an extinction event occurred in our oceans, it will greatly affect humans since much of our food supply is reliant on fish or other marine animals.

Ocean acidification due to global warming will also change the reproductive cycles of reef fish who normally spawn during late spring and fall. On top of this, there will be increased mortality rates among the larvae of coral reef fish since the acidic environment slows down their development. The hypothalamo-pituitary-gonadal (HPG) axis is one of the regulatory sequences in fish for reproduction, which is mainly controlled by surrounding water temperature. Once a minimum temperature threshold is reached, the production of hormone synthesis increases significantly, causing the fish to produce mature egg and sperm cells. Spawning in the spring will have a shortened period, while fall spawning will be delayed substantially. Because of the increased CO_2 levels in the ocean from coral bleaching, there will be a substantial decrease in the number of young reef fish that survive to maturity. There is also evidence that shows that embryo and larval stage fish have not matured enough to express the appropriate levels of acid/base regulation that is present in adults. These will ultimately lead to hypoxia due to the Bohr effect driving oxygen off of hemoglobin. This will lead to increased mortality as well as impaired growth performance for fish in slightly acidic conditions relative to the normal proportion of acid dissolved in marine water.

In addition, ocean acidification will make fish larvae more sensitive to the surrounding pH since they are more sensitive to environmental fluctuations than adults. In addition, larvae of common prey species will have lower survival rates, which in turn will eventually cause the species to become endangered or extinct. Also, elevated CO_2 in marine environments can lead to neurotransmitter interference in both predator and prey fish which increases their mortality rate. It has also been shown that when fish spend considerable time in high concentrations of dissolved CO_2 up to 50,000 micro-atmospheres (μatm) of CO_2 in marine environments, cardiac failure leading to death is much more common than in normal CO_2 environments. In addition, fish that live in high CO_2 environments are required to spend more of their energy to keep their acid/base regulation in check. This diverts precious energy resources from important parts of their life cycle such as feeding and mating to keep their osmoregulatory functions in check.

Another important consequence of ocean acidification is that endangered species will have fewer places where their eggs are laid. For species with poor larval dispersal, it puts them at a greater risk of extinction because natural egg predators will find their nests or hiding places and eat the next generation.

Other Biological Impacts

Aside from the slowing and reversing of calcification, organisms may suffer other adverse effects, either indirectly through negative impacts on food resources, or directly as reproductive or physiological effects. For example, the elevated oceanic levels of CO2 may produce CO_2-induced acidification of body fluids, known as hypercapnia. Also, increasing ocean acidity is believed to have a range of direct consequences. For example, increasing acidity has been observed to: reduce metabolic rates in jumbo squid; depress the immune responses of blue mussels; and make it harder for juvenile clownfish to tell apart the smells of non-predators and predators, or hear the sounds of their predators. This is possibly because ocean acidification may alter the acoustic properties of seawater, allowing sound to propagate further, and increasing ocean noise. This impacts all animals that use sound for echolocation or communication. Atlantic longfin squid eggs took longer to hatch in acidified water, and the squid's statolith was smaller and malformed in animals placed in sea water with a lower pH. The lower PH was simulated with 20–30 times the normal amount of CO_2. However, as with calcification, as yet there is not a full understanding of these processes in marine organisms or ecosystems.

Another possible effect would be an increase in red tide events, which could contribute to the accumulation of toxins (domoic acid, brevetoxin, saxitoxin) in small organisms such as anchovies and shellfish, in turn increasing occurrences of amnesic shellfish poisoning, neurotoxic shellfish poisoning and paralytic shellfish poisoning.

Although red tide is harmful, other beneficial photosynthetic organisms may benefit from increased levels of carbon dioxide. Most importantly, seagrasses will benefit. An experiment done in 2018 concluded that as seagrasses increased their photosynthetic activity, calcifying algae's calcification rates rose. This could be a potential mitigation technique in the face of increasing acidity.

Ecosystem Impacts Amplified by Ocean Warming and Deoxygenation

While the full implications of elevated CO_2 on marine ecosystems are still being documented, there is a substantial body of research showing that a combination of ocean acidification and elevated ocean temperature, driven mainly by CO_2 and other greenhouse gas emissions, have a compounded effect on marine life and the ocean environment. This effect far exceeds the individual harmful impact of either. In addition, ocean warming exacerbates ocean deoxygenation, which is an additional stressor on marine organisms, by increasing ocean stratification, through density and solubility effects, thus limiting nutrients, while at the same time increasing metabolic demand.

Meta analyses have quantified the direction and magnitude of the harmful effects of ocean acidification, warming and deoxygenation on the ocean. These meta-analyses have been further tested by mesocosm studies that simulated the interaction of these stressors and found a catastrophic effect on the marine food web, i.e. that the increases in consumption from thermal stress more than negates any primary producer to herbivore increase from elevated CO_2.

Nonbiological Impacts

Leaving aside direct biological effects, it is expected that ocean acidification in the future will lead to a significant decrease in the burial of carbonate sediments for several centuries, and even the dissolution of existing carbonate sediments. This will cause an elevation of ocean alkalinity, leading to the enhancement of the ocean as a reservoir for CO_2 with implications for climate change as more CO_2 leaves the atmosphere for the ocean.

Impact on Human Industry

The threat of acidification includes a decline in commercial fisheries and in the Arctic tourism industry and economy. Commercial fisheries are threatened because acidification harms calcifying organisms which form the base of the Arctic food webs.

Pteropods and brittle stars both form the base of the Arctic food webs and are both seriously damaged from acidification. Pteropods shells dissolve with increasing acidification and the brittle stars lose muscle mass when re-growing appendages. For pteropods to create shells they require aragonite which is produced through carbonate ions and dissolved calcium. Pteropods are severely affected because increasing acidification levels have steadily decreased the amount of water supersaturated with carbonate which is needed for aragonite creation. Arctic waters are changing so rapidly that they will become undersaturated with aragonite as early as 2016. Additionally the brittle star's eggs die within a few days when exposed to expected conditions resulting from Arctic acidification. Acidification threatens to destroy Arctic food webs from the base up. Arctic food webs are considered simple, meaning there are few steps in the food chain from small organisms to larger predators. For example, pteropods are "a key prey item of a number of higher predators – larger plankton, fish, seabirds, whales". Both pteropods and sea stars serve as a substantial food source and their removal from the simple food web would pose a serious threat to the whole ecosystem. The effects on the calcifying organisms at the base of the food webs could potentially destroy fisheries. The value of fish caught from US commercial fisheries in 2007 was valued at $3.8 billion and of that 73% was derived from calcifiers and their direct predators. Other organisms are directly harmed as a result of acidification. For example, decrease in the growth of marine calcifiers such as the American lobster, ocean quahog, and scallops means there is less shellfish meat available for sale and consumption. Red king crab fisheries are also at a serious threat because crabs are calcifiers and rely on carbonate ions for shell development. Baby red king crab when exposed to increased acidification levels experienced 100% mortality after 95 days. In 2006, red king crab accounted for 23% of the total guideline harvest levels and a serious decline in red crab population would threaten the crab harvesting industry. Several ocean goods and services are likely to be undermined by future ocean acidification potentially affecting the livelihoods of some 400 to 800 million people depending upon the emission scenario.

Impact on Indigenous Peoples

Acidification could damage the Arctic tourism economy and affect the way of life of indigenous peoples. A major pillar of Arctic tourism is the sport fishing and hunting industry. The sport fishing industry is threatened by collapsing food webs which provide food for the prized fish. A decline in tourism lowers revenue input in the area, and threatens the economies that are increasingly dependent on tourism. The rapid decrease or disappearance of marine life could also affect the diet of Indigenous peoples.

Possible Responses

Demonstrator calling for action against ocean acidification at the People's Climate March.

Reducing CO_2 Emissions

Members of the InterAcademy Panel recommended that by 2050, global anthropogenic CO_2 emissions be reduced less than 50% of the 1990 level. The 2009 statement also called on world leaders to:

- Acknowledge that ocean acidification is a direct and real consequence of increasing atmospheric CO_2 concentrations, is already having an effect at current concentrations, and is likely to cause grave harm to important marine ecosystems as CO_2 concentrations reach 450 parts-per-million (ppm) and above.
- Recognize that reducing the build up of CO_2 in the atmosphere is the only practicable solution to mitigating ocean acidification.
- Reinvigorate action to reduce stressors, such as overfishing and pollution, on marine ecosystems to increase resilience to ocean acidification.

Stabilizing atmospheric CO_2 concentrations at 450 ppm would require near-term emissions reductions, with steeper reductions over time.

The German Advisory Council on Global Change stated:

> "In order to prevent disruption of the calcification of marine organisms and the resultant risk of fundamentally altering marine food webs, the following guard rail should be obeyed: the pH of near surface waters should not drop more than 0.2 units below the pre-industrial average value in any larger ocean region (nor in the global mean)."

One policy target related to ocean acidity is the magnitude of future global warming. Parties to the United Nations Framework Convention on Climate Change (UNFCCC) adopted a target of limiting warming to below 2 °C, relative to the pre-industrial level. Meeting this target would require substantial reductions in anthropogenic CO_2 emissions.

Limiting global warming to below 2 °C would imply a reduction in surface ocean pH of 0.16 from pre-industrial levels. This would represent a substantial decline in surface ocean pH.

On 25 September 2015, USEPA denied a 30 June 2015, citizens petition that asked EPA to regulate CO_2 under TSCA in order to mitigate ocean acidification. In the denial, EPA said that risks from ocean acidification were being "more efficiently and effectively addressed" under domestic actions, e.g., under the Presidential Climate Action Plan, and that multiple avenues are being pursued to work with and in other nations to reduce emissions and deforestation and promote clean energy and energy efficiency.

On 28 March 2017 the US by executive order rescinded the Climate Action Plan. On 1 June 2017 it was announced the US would withdraw from the Paris accords, and on

12 June 2017 that the US would abstain from the G7 Climate Change Pledge, two major international efforts to reduce CO_2 emissions.

Geoengineering

Geoengineering has been proposed as a possible response to ocean acidification. The IAP statement said more research is needed to prove that this would be safe, affordable and worthwhile:

> Mitigation approaches such as adding chemicals to counter the effects of acidification are likely to be expensive, only partly effective and only at a very local scale, and may pose additional unanticipated risks to the marine environment. There has been very little research on the feasibility and impacts of these approaches. Substantial research is needed before these techniques could be applied.

Reports by the WGBU, the UK's Royal Society , and the US National Research Council warned of the potential risks and difficulties associated with climate engineering.

Iron Fertilization

Iron fertilization of the ocean could stimulate photosynthesis in phytoplankton. The phytoplankton would convert the ocean's dissolved carbon dioxide into carbohydrate and oxygen gas, some of which would sink into the deeper ocean before oxidizing. More than a dozen open-sea experiments confirmed that adding iron to the ocean increases photosynthesis in phytoplankton by up to 30 times. While this approach has been proposed as a potential solution to the ocean acidification problem, mitigation of surface ocean acidification might increase acidification in the less-inhabited deep ocean.

A report by the UK's Royal Society reviewed the approach for effectiveness, affordability, timeliness and safety. The rating for affordability was "medium", or "not expected to be very cost-effective". For the other three criteria, the ratings ranged from "low" to "very low" (i.e., not good). For example, in regards to safety, the report found a "high potential for undesirable ecological side effects", and that ocean fertilization "may increase anoxic regions of ocean ('dead zones')".

Impact of Oil Spills on Marine Life

Environmental pollution caused by petroleum is of great concern. This is because petroleum hydrocarbons are toxic to all forms of life and harm both aquatic and terrestrial ecosystems. The pollution of marine habitats has caught the attention of researchers and environmentalists. This is due to the serious impact of oil spills on marine life, as well as on people whose career relies on the exploitation of the sea's resources. Additionally, marine life may be affected by clean-up operations. It may also be indirectly affected by the physical damage to the habitats in which plants and animals live in.

Petroleum marine fuel spills, which result from damage, transportation accidents and various other industrial and mining activities, are classified as hazardous waste. They are considered to be the most frequent organic pollutants of aquatic ecosystems.

In recent years, there have been numerous studies regarding the levels of contamination of the seawater by hydrocarbons. The majority of these studies were conducted following the Gulf War of 1991and after, the BP Deepwater Horizon (DWH) oil spill on 20th April, 2010.

Major Oil Spills Related to Marine Environments

Saltwater bodies are referred to as 'marine environments', with the oceans covering about 70% of the earth. Statistics estimate that 3.2 million tonnes of oil per year are released from all sources into the environment. The majority of this oil is due to general shipping and industrial activities. During the Iran-Iraq war, approximately 2 million barrels of oil were discharged into the Arabian Gulf sea water. These included 1.5 million barrels from the Nawruz blow-out in 1983. Following the Gulf War in 1991, between 4 and 8 million barrels (1,000 tonnes = 7,500 barrels) were released into the Gulf and the Kuwaiti Desert, making this the largest oil spill in history at that time. When compared to other major spills, the size of this spill attracted global attention. For example, the Amoco Cadiz off the coast of Brittany , spilling 200,000 tonnes (1.5 million barrels); the Torrey Canyon, Braer, Sea Empress and the super tanker Breaf off the coast of Shetland in 1993, spilling a maximum of 84,000 tonnes (607,300 barrels); the Exxon Valdez in Prince William Sound or Alaska , which was approximately 36,224 tonnes (261,904 barrels). Of the recorded oil spills, more than 80% were less than 1,000 tonnes (7,500 barrels) and only 5% were greater than 10,000 tonnes. An accepted average sample size of an oil spill is about 700 tonnes (5,061 barrels). Previous observations indicated that the number of large spills (>700 tonnes) has decreased significantly over the last 30 years.

Table: Number of Oil Spills over 7 Tonnes from 1970 to 2013.

Year	7-700 tonnes	>700 tonnes	Quantity Spilled x 103 tonnes
1970-1974	189	125	1114
1975-1979	342	117	2012
1980-1984	221	41	570
1985-1989	124	48	513

1990-1994	165	48	907
1995-1999	108	25	194
2000-2009 2010-2013	182 28	- -	213 22

During the 1990s, the average number of large spills per year was about a third of the amount that was witnessed during the 1970s. When looking at the quantities of oil spilled, it should be noted that 1, 133, 000, 213, 000 tonnes of oil were lost in the 1990s and 2000s, respectively. In a four-year period, 2010-2013, there were 22,000 tonnes of oil lost. Table shows a brief summary of the top 22 major oil spills since 1967.

Table: Top 22 Major Oil Spills since 1967.

Ship name	Year	Location	Spill (103) tonnes
Torrey Canyon	1967	Scilly Isles, UK	119
Sea Star	1972	Gulf of Oman	115
Jakob Maersk	1975	Oporto, Portugal	88
Urquiola	1976	La Coruna, Spain	100
Hawaiian Patriot	1977	300 nautical. miles off Honolulu	95
Amoco Cadiz	1978	Off Brittany, France	223
Atlantic Empress	1979	Off Tobago, West Indies	287
Independenta	1979	Bosphorus, Turkey	94
Irenes Serenade	1980	Navarino Bay, Greece	100
Castillo de Bellver	1983	Off Saldanha Bay, South Africa	252
Novo	1985	Off Kharg Island, Iran	70
Odyssey	1988	700 nautical. miles off Nova Scotia, Canada	132
Khark 5	1989	120 nautical. miles off Atlantic coast of Morocco	70
Exxon Valdez	1989	Prince William Sound, Alaska, USA	37
ABT Summer	1991	700 nautical miles off Angola	260
Haven	1991	Genoa, Italy	144
Aegean Sea	1992	La Coruna, Spain	74
Katina P.	1992	Off Maputo, Mozambique	67
Braer	1993	Shetland Islands, UK	85
Sea Empress	1996	Milford Haven, UK	72
Prestige	2002	Off Galicia, Spain	63
Hebel Spirit	2007	South Korea	11

The BP Deepwater Horizon (DWH) oil spill on 20th April, 2010, initiated the discharge of more than 2.6 million gallons (over 800 million litres) of oil into the Gulf of Mexico over approximately three months. This oil spill was the second largest in human history. During the 1991 Gulf War, the deliberate release of over 6 million barrels of oil into the marine environment was considered as the largest in history.

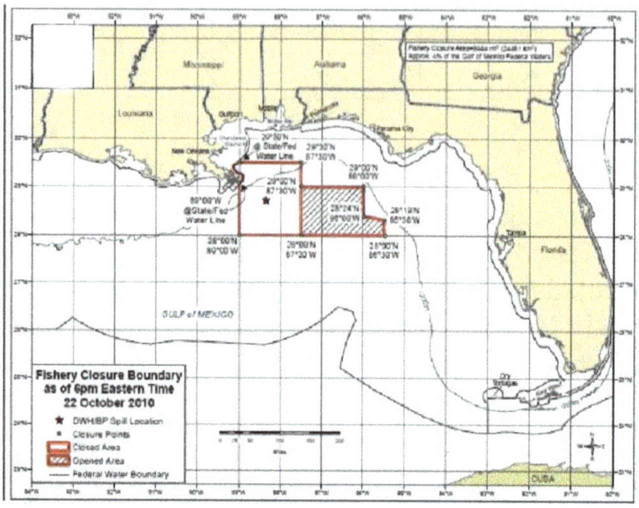

The Site of the Gulf Spill: Closed Areas as of 22nd October.

Crude Oil and its Properties

Crude oil is a complex mixture of organic compounds. These mainly consist of hydrocarbons, in addition to heterocyclic compounds and some heavy metals. The different hydrocarbons that make up crude oil come in a wide range of molecular weights and structure compounds. These compounds include methane gas, high molecular weight tars, asphaltenes, resins, waxes and bitumens. They also include straight and branched chains, single or condensed rings and aromatic rings such as the monocyclic (benzene, toluene, ethylbenzene and xylene). They additionally include polycyclic aromatic hydrocarbons (PAHs) such as naphthalene, anthracene and phenanthrene. Examples of the chemical structure of some common components of crude petroleum are shown in figure.

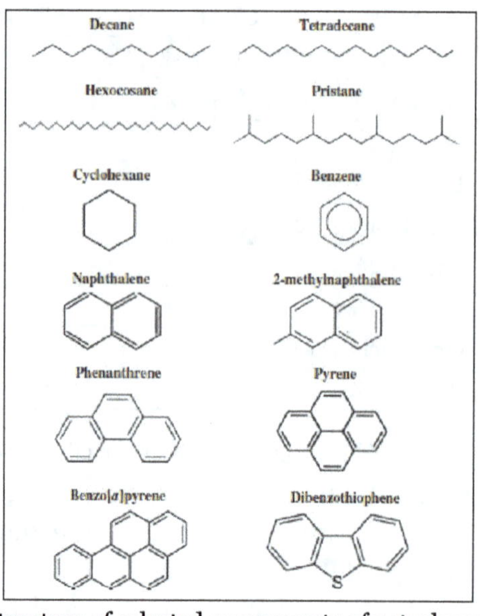

Structure of selected components of petroleum.

Toxicity of Oil

The general effects of oil toxicity depend on a multitude of factors. These include the oil composition and characteristics (physical and chemical), condition (i.e., weathered or not), exposure routes and regimen, and the bioavailability of the oil. One major effect of oil is narcosis, a reversible anaesthetic effect caused by the oil partitioning into the cell membrane and nervous tissue. This causes dysfunctions of the central nervous system.

The additive toxic effect of hydrocarbons can lead to mortality, if the levels exceed the threshold concentration. When oil hydrocarbons are ingested by marine animals, they travel to the liver where enzymes activate PAHs to become more toxic and reactive products. The metabolites of polycyclic aromatic hydrocarbons (PAHs) and aliphatic hydrocarbons can be highly toxic and carcinogenic. In particular, PAHs are the major contributors to toxicity, with different metabolic pathways producing metabolites. These have oxidative and carcinogenic properties due to their ability to attack and bind to DNA and proteins. Hydrocarbons have a volatile nature and, therefore, inhalation of them results in respiratory tract irritation and narcosis of mammals and birds.

Physical contact is the major route of exposure and usually affects birds and furred mammals. These animals rely on their outer coats for buoyancy and warmth. Consequently, they often succumb to hypothermia, drowning and smothering when oil flattens and adheres to the outer layer. A second general exposure route is through the ingestion or inhalation of the hydrocarbon by organisms that reside on the surface. Exposure by these routes leads to absorption into the bloodstream via the gastrointestinal or respiratory tracts.

Toxicity of Oil Dispersants

Oil dispersants (57 chemical ingredients approved for use by the US EPA) are a common tool used after oil spills in marine environments. They break up oil slicks on the water surface and increase the oil's rate of biodegradation. Oil dispersants are quickly used when other means, such as oil containment and removal, are insufficient. However, consequences of the toxicity of oil spill dispersants alone or in the presence of oil must be evaluated. Generally, undispersed oil poses the greatest threat to shorelines and surface dwelling organisms. However, most dispersed oil remains in the water column where it mainly threatens pelagic and benthic organisms.

Several studies have compared the toxicity of oil spill dispersants alone or in the presence of oil. Analyses of tests conducted on a variety of species of aquatic life showed that crustaceans are more sensitive to oil dispersant exposure, compared with fish. A study by indicated that the species with the least amount of protective shell or external tissue is the most sensitive to oil dispersant exposure.

It has been shown that the use of oil dispersants increases the exposure and uptake of PAHs by fish. This is particularly the case with fish that live throughout the water column of coastal areas, the ocean and lakes. Researchers found that 'the risk of PAH toxicity especially to sensitive life stages, such as eggs and larvae, is enhanced by chemical dispersion'. In addition, 'concentrations of LMWPAHs and HMWPAHs (low and high molecular weight PAHs) were found to be higher in the water column following the application of chemical dispersants to the surface slicks'. They reported that chemical dispersants mobilize PAHs to toxic concentrations as the biomarker ethoxyresorufin-O-deethylase (EROD) activity is increased after exposing newly-hatched mummichog (Fundulusheteroclitus) for 96 h to crude oil and chemically dispersed crude oil.

The most sensitive species was the snakelocks anemone Anemonia viridis with a 48-hour LOEC of 20 ppm (nominal). This was followed by mussel (Mytilus edulis) feeding rate (50 ppm), seagrass (Zostera marina) photo synthetic index (80 ppm), burrowing amphipod (Corophium volutator) (mortality 175 ppm) and mussel lethality (250 ppm).

Moreover, in a study, adults of four species of wild-caught Newfoundland nearshore fishes were exposed for four days in flow-thru conditions to the dispersant Corexit 9527 alone, water accommodated fraction (WAF) of Hibernia light crude oil alone and dispersed Hibernia crude oil. Test toxicants (20 to 50 ml) were added daily to 300 L tanks for four days, followed by up to six weeks in clean water. The investigators did not report exposure temperatures or toxicant concentrations except to note that initial daily concentrations were 50-100 ppb for Hibernia water-accommodated fraction (WAF). On the first day, the caplein responded to the dispersant by swimming erratically. On the second and subsequent days, they responded by death accompanied by hemorrhaging of the gill lamellae.

Fate of Oil Spills in the Marine Habitats

After oil is spilled at sea and with the effect of wind and water current, the oil spreads out and moves on the water surface as a slick a few millimetres thick. At the same time, it undergoes a series of chemical and physical changes. These processes are collectively termed 'weathering'. Weathering causes the spilled oil to break down and become heavier than water. Some of these processes, like the natural dispersion of oil into water, lead to the removal of the oil from the sea surface and facilitate its natural breakdown in the marine environment. Others, particularly the formation of water-in-oil emulsions, cause the oil to become more persistent and remain at sea or on the shoreline for prolonged periods of time. The speed and relative importance of these processes depend on a number of factors. These include the quantity spilled, the oil's initial physical and chemical characteristics, weather and sea conditions and whether the oil remains at sea or is washed ashore. Ultimately, the marine environment usually eliminates spilled oil through the long-term process of biodegradation.

Impact of Oil Spills on Marine Organisms

Ultimately, the impact of oil on marine organisms depends on the fate of the oil. As previously described, when oil is present in the environment, it is either dispersed in the top layer of the water (littoral zone) or remains on the surface and, consequently, on the coastal areas. If the oil is not dispersed, it remains on the surface. In this case, currents bring the oil towards coastal areas which harms coastal organisms like invertebrates, mammals and birds. However, if the oil is dispersed, organisms, such as fish, plankton and larvae, are immediately subjected to oil toxicity.

Impact of Oil Spills on Planktonic Organisms

Zooplankton is a particularly important food resource, especially for baleen whales. It can influence or control the primary productivity by top-down effectsin return. Its population dynamic change can influence the biomass of other marine animals like fish by bottom-up effects. Some zooplankton, such as copepods, euphausiids and mysids, assimilate hydrocarbons directly from seawater and by ingesting oil droplets and oil contaminated food.The ingestion of oil by these organisms often causes mortality, while surviving organisms often show developmental and reproductive abnormalities.

The impact of petroleum pollution on marine plankton has been a great cause for concern. Generally, oil WAF toxicity enhances with increasing carbonic chain length and benzene ring number.

For the oil WAF, reported that plankton are capable of accumulating PAH due to their great lipophilic abilities. They, therefore, stimulate various harmful effects. The investigators reported that marine plankton is highly sensitive to the petroleum WAF, as the order of median effective/lethal concentration is as low as lg/L or mg/L. Examinations of the toxicity effect of 10 polycyclic aromatic hydrocarbons associated with the Prestige fuel oil spill on adult copepods (Oithona davisae) revealed that the PAHs had narcotic effects on these organisms.

Impact of Oil Spills on Benthic Organisms and Invertebrates

Benthic invertebrates and higher forms, such as the sand eel and Ammodvtesamericanus (a main food resource of Atlantic humpback whales), may accumulate petroleum hydrocarbons from water, contaminated sediments and food. Thus, these whales are adversely affected.

Bivalve molluscs tend to accumulate petroleum hydrocarbons to higher concentrations and retain them longer than other taxa. This is essentially due to the lack of a mixed function oxygenase (MFO) system that makes them unable to metabolize the compounds to execrable polar metabolites. Thus, they are likely to transfer them to their predators. Marine mammals that rely heavily on bivalve molluscs for food, such as the walrus and otter, share a higher risk of ingesting petroleum hydrocarbons. Benthic amphipods are quite sensitive to spilled oil. They are among the first marine animals killed and the slowest to recover. However, most marine crustaceans have a well-developed

MFO system. As a result, they are able to metabolize and excrete accumulated hydrocarbons quite rapidly.

Previous studies have explored the recovery of the invertebrate populations after oil spills. In the intertidal habitat, the biological recovery of the exposed shores is faster than the sheltered shores. This is because strong wave action promotes the removal of contamination and the animals and plants of exposed shores tend to be severely affected. Thus, they are better able to re-colonize an impacted shore quickly. Sublittoral habitats are generally contaminated by sedimentation of oiled particulate material and clean-up for these habitats is not practiced. As a result, the recovery of subtidal communities impacted by oil spills usually takes a longer time. For example, the Abra alba bivalve sand community in the Bay of Morlaix, Brittany was severely affected by the Amoco Cadiz oil spill. After the spill in 1978, the biomass values for the sand community immediately fell. However, within two years, they recovered to pre-spill levels. Productivity also showed similar trends.

Invertebrate populations, such as the amphipod sand hopper, Ampelisca, are quite sensitive to oil pollution and, for various reasons, are slow to re-populate and 'recover'. For example, the initial impact of the spill in the Bay of Morlaix, following the Amoco Cadiz spill in 1978, was to kill off populations of the amphipod sand hopper, Ampelisca, which dominated the community. Although the sediment was rapidly purged of the contaminating oil, it has been noted that Ampelisca was back to its pre-spill population density after 10 years. The standing crop biomass and productivity was restored much more rapidly as the ecological niches had been occupied by other opportunists like the bivalve Abra and the worm Nephtys. These had quickly filled the place left by the Ampelisca.

In other studies on oil spill accidents, it has been shown that six months after the Prestige oil spill and clean-up campaign, invertebrate populations of the exposed sandy beaches, notably the isopod Eurydice, the spionid polychaete Scoleleoius squamata, nemerteans and Diptera, were significantly reduced. Furthermore, their abundance inversely related with the oil pollution gradient. The number of taxa was reduced but not the diversity values. The only clam on these sandy beaches was Donax. Before the spill and clean-up, it occurred in six beaches but afterwards, only in one. Upper dry sand communities were particularly reduced due to both oil toxicity and extensive beach grooming that also removed seaweed wrack.

Marine invertebrates are unique living organisms in different aspects. They have been explored as a model for several biological markers as a result of oil spill pollution. The mussel response was assessed using a set of biomarkers (acetylcholinesterase (AChE), glutathione S-transferase (GST), catalase (CAT), malondialdehyde (MDA) and deoxyribonucleic acid (DNA) adducts, related to the metabolism of the organic contaminants. The results of a series of validation tests revealed that there were no significant reductions in the GST or CAT levels. Six months immediately following the accident, observations indicated that the DNA adducts and MDA levels were high and the levels of AChE were significantly lower during the first year of the survey. This suggested a general stress. A simple multivariate graphic method - the integrated biomarker response index - was used to combine four of the five validated biomarkers and quantify the degree of impact on mussels at different sites. The results indicated that mussel populations were affected by the oil spill during only the first year after the accident.

Impact of Oil Spills on Coral Reefs

In addition, recreational attractions for divers, coral reefs are considered to be important constituents of marine ecosystems. This is because they are important nurseries for shrimp, fish and other animals. The aquatic organisms that live within and around the coral reefs are at risk of exposure to the toxic substances within oil, as well as smothering. They are rapidly deteriorating because of a variety of environmental and anthropogenic pressures. Thus, they are suffering significant changes in diversity, species abundance and habitat structure worldwide.

Oil dispersants are potentially harmful to marine life including coral reefs. In a study using coral nubbins in coral reef ecotoxicology testing, found that dispersed oil and oil dispersants are harmful to soft and hard coral species at early life stages. The investigators also employed a 'nubbin assay' on more than 10, 000 coral fragments to evaluate the short- and long-term impacts of dispersed oil fractions (DOFs) from six commercial dispersants (Slickgone, Petrotech, Inipol, Biorieco, Emulgal and Dispolen) and the dispersants and water-soluble-fractions (WSFs) of Egyptian crude oil on two Indo-Pacific branching coral species, Stylophora pistillata and Pocillopora damicornis. They found that the dispersant concentrations recommended by the manufacturer were highly toxic and resulted in mortality of all nubbins. The dispersed oil and the dispersants were significantly more toxic than the crude oil WSFs. As corals are very sensitive to oil detergents and dispersed oil, the results of these assays indicated the negative and harmful effect of using any oil dispersant in coral reefs and in the area closely around them. These dispersants were rated based on their eco-toxicological impacts on the corals. Observations indicated a scale from the least to the most harmful agent, as follows:

Slickgone > Petrotech > Inipol > Biorieco > Emulgal > Dispolen.

Impact of Oil Spills on Fish

Due to the well-developed hepatic mixed function oxygenase (MFO) system, in addition to the reactivity of the metabolites that would not be released in a toxic form during digestion and absorption, most fish, even in heavily oil-contaminated environments, do not accumulate and retain high concentrations of petroleum hydrocarbons. Thus, they are not likely to transfer them to predators. Thus, no serious threat is predicted.

Generally, marine carnivores are inefficient assimilators of petroleum compounds in food. For this reason, and because all prey species are able to release hydrocarbons from their tissues, the marine food chain biomagnification does not occur. Thus, there is an indirect correlation between a marine mammal's trophic level and the concentration of residues that it might consume. In fact, as top carnivores that feed on large pelagic fish and seals, polar bears and killer whales are less likely to be exposed to petroleum in their food than other species, such as walrus and baleen whales, which feed on zooplankton and benthic invertebrates.

Chronic exposure of juvenile pink salmon (Oncorhynchus gorbuscha) to Alaska North slope crude oil resulted in a variety of responses. These included melanosis, erratic swimming, loss of equilibrium, reduced mobility and startled response. Interestingly, the effects were not enhanced by the phototoxicity from UV irradiation, presumably of the highly pigmented nature of the fish's skin (in contrast to phototoxicity in translucent early life stages of marine invertebrates). When they examined the phototoxicity of the water accommodated fractions of weathered crude oil, they found that pink salmon may be at less risk of photoenhanced toxicity, compared to the more translucent early-life stages of several other Alaskan species.

Impact of Oil Spills on Seabirds

As one of the major routes of exposure, physical contact usually affects birds. For example, thousands of African penguins (Spheniscus demerus) were oiled following the 2000 Treasure oil spill in South Africa.

An evaluation of the impact of oil spills on seabirds has not been fully appreciated during incidents, despite pressure from the public concern, media and other interested parties for precise and up-to-date information on the damage. Consequently, the approximate numbers of seabird casualties involved in many major spills have only been estimated, while impacts at the population level have been difficult to determine. Natural variation and the huge range of factors that influence bird population statistics make it difficult to assess the impact of oil spill on sea birds.

The relation between the size of an oil spill and the number of seabird casualties is not directly proportional. Moreover, estimates of the number of seabird deaths from oil slicks are highly speculative. This is because an unknown number of oiled birds may die at sea and not reach the coast. For example, following the Exxon Valdez oil spill, over 35,000 seabird carcasses were recovered in the northern Gulf of Alaska. However, after the Braer accident, even though the Braer spill (85,000 tonnes) was almost two and a half times as large as that of the Exxon Valdez, only 1,500 dead birds were counted.

Researchers argued that the rapid recovery of the murre breeding colonies in Alaska and the number of dead birds might be lower than was estimated. They also suggested that breeding pairs may have been replaced by younger birds that came in from the open sea. On the other hand, argued that the impacts of the spill may have been diffused over a large area, permitting local recovery and making it difficult to detect any changes in local abundance or habitat occupancy. Nine years after the Exxon Valdez oil spill, in their long-term effects observation, claimed that the populations of most bird species have not recovered and others still show potential population effects. However, the report is at variance with other findings.

During the winter season of 2003, approximately 64,000 tonnes of Prestige heavy fuel oil spilled off Galacia, Spain. As a result, 10 % of the European shag (cormorants, Phalacrocorax aristotellid) were killed. In turn, this resulted in a 50% reduction in the 2003 breeding success of oiled colonies, compared to the unoiled colonies. The kidneys of 32 oiled shag had extensive tissue damage and haemorrhaging. They further found that the oiled shag were heavily infected with an eimeriorin Apicomplex coccidian, which under normal circumstances is not pathogenic.

Table: Comparison of the number of dead seabirds recovered following the exxon valdez and braer oil spills.

Species group	Alaskan spill	Shetland spill
Sea ducks (eiders, etc.)	1440	167
Mergansers	121	1
Loons	395	14
Grebes	462	0
Heron	1	3
Geese/swans	9	0
Gulls	696	74
Kittiwakes	1225	133
Cormorants/shags	836	864
Shearwaters	3399	0
Fulmars	870	31
Guillemots/murres	20562	220
Other auks	2174	29
Bald eagles	125	NA
Other birds	3152	0
Total	35467	1536

Impact of Oil Spills on Marine Mammals

Marine mammals include bottlenose dolphins, fins, humpbacks, rights, sei whales, sperm whales, manatees, cetaceans, seals, sea otters and pinnipeds. As previously indicated, the physical contact of oil with furred mammals usually affects these animals because they rely on their outer coats for buoyancy and warmth. Consequently, these animals often succumb to hypothermia, drowning and smothering when oil flattens and adheres to the outer layer.

As part of their activities, all marine mammals spend a considerable amount of time at the surface. Here, they swim, breathe, feed or rest. Thus, the possibility of their contact with a surface slick, water-in-oil emulsion, or tar balls, is high. In heavy pelage marine mammals, such as fur seals, sea otters and polar bears, this contact may lead to fouling. Polar bears and otters groom themselves regularly as a means of maintaining the insulating properties of the fur and may, thereby, ingest oil. Animals with smooth surfaces or relatively little to no pelage, such as whales, dolphins, manatees and most seals, have an advantage as oil would have fewer tendencies to adhere to their surface.

Some baleen whales are skim-feeders, i.e., they eat at the surface. When in an area of slick or tar balls, this behavior can lead to foul in the feeding apparatus. Tarry residues, in particular, can coat the baleen plates.

Animals, such as narwhals, belugas, ringed seals, walruses and polar bears in Polar Regions, spend most of their time at the ice edge in leads, polynyas and breathing holes. This is where spilled oil tends to accumulate.

Oil that contaminates a shore is likely to severely affect pinnipeds. Pinnipeds require such areas for nursery and, to a lesser extent, otters and bears. Some of the oil is eventually returned in subtidal sediments, where it may transfer to gray whales, walrus and some seals. Such species feed heavily on benthic animals.

When marine mammals encounter fresh oil, they are likely to inhale volatile hydrocarbons evaporating from the surface slick. These volatile fractions contain toxic monoaromatic hydrocarbons (benzene, toluene and xylenes) and low molecular weight aliphatics with anaesthetic properties. The inhalation of these volatile hydrocarbon compounds is potentially harmful. The inhalation of concentrated petroleum vapours can cause the inflammation of and damage to the mucus membranes of airways, lung congestion or even pneumonia. Volatile benzene and toluene, which can be inhaled, can be transferred rapidly from the bloodstream into the lungs. Furthermore, they can accumulate from the blood into the brain and liver, causing neurological disorders and liver damage.

Marine mammals are probably poor accumulators to oil directly from the solution or dispersion in the water column. This is because the skin of cetaceans is relatively impermeable to oil. Additionally, most marine mammals do not drink large volumes of seawater. Thus, a significant accumulation of hydrocarbons by this route is unlikely to occur.

There is an extensive and diverse database regarding subjects that deal with effects of oil on marine mammals and those aspects of an animal's life history vulnerable to exposure of spilled oil. This database is summarized in a publication by the Department of Interior/Minerals Management Servic /Atlantic OCS region, Canada. Described the effects of oil on marine mammal populations. At the same time,described the physiological and toxicological effects of oil on each of the marine mammal groups.

Impact of Oil Spills on Marine Plants

In several aspects, aquatic plants are important to the functioning of ecosystems. These include the fact that they are oxygen producers, their ability to sequester carbon and for their base position in aquatic food chains. In addition, they serve as nursery, feeding and breeding habitats for a variety of animal and plant species, including recreationally and commercially important fish.

Plants and animals are affected by the oil in which they come into contact with as a result of an oil spill. Dispersants and dispersed oils to algae and aquatic plants, summarized the reported phytotoxicities of oils, dispersants and their combinations to aquatic plants. They assessed the ability of the reviewed database to support toxicity predictions and evidence-based risk assessments. The phytotoxicity database mainly includes the results of research conducted after oil spills to marine waters. The toxicity of at least 41 crude oils and 56 dispersants were recorded. At least 107 response parameters were monitored for 85 species of unicellular and multicellular algae, 28 wetland plants, 13 mangroves and nine seagrasses. Due to experimental diversity, the effect concentrations available from this toxicity database are varied and diverse. As a result, there are restricted phytotoxicity predictions and identification of sensitive species, life stages and response parameters. Thus, the impact of toxicity of petrochemicals and dispersants on aquatic plants was not supported by this database.

The effects of heavy fuel oil contamination on the growth and the development of Salicornia fragilis Ball and Tutin, a salt-marsh edible species, were studied as greenhouse experiments by. Phytotoxicity assessments and PAH shoots assays were followed to measure the impact of petroleum on plant development. Syptoms like chlorosis, yellowing, growth reduction and perturbations in developmental parameters were visually observed. Shoot coating appeared to be less than through soil and the investigators observed more marked effects on plants as an indication of the degree of pollution. However, a significant bioaccumulation of PAHs in shoot tissues was also found, even at very low levels of contamination. These highly related to the conditions of exposure to oil. This strong relationship between the PAH contents of Salicornia plants and growth reduction suggest a chemical toxicity of fuel oil. The investigators concluded that the type and degree of fuel oil contamination are two important factors in controlling the impact of fuel oil on S. fragilis.

In 1986, more than 8 million litres of crude oil spilled into a complex region of mangroves, seagrasses and coral reefs, just east of the Caribbean entrance to the Panama Canal. Intertidal mangroves, sea grasses, algae and associated invertebrates were covered by oil and died soon after. Investigators reported that seedlings of red mangrove, Rhizophoram angle, which were transplanted to heavily oiled sites, did not produce new leaves. This contrasted the transplants at an unoiled site. Entire beds of intertidal seagrass Thalassiat estudinum were killed on some heavily oiled reef flats, as shown by the abundant oil-covered dead leaves washed ashore, as well as the dead, but intact, root-rhizome mats. In contrast, the subtidal Thalassia survived everywhere after the spill. Having said this, in the heavily oiled areas, the leaves of the subtidal Thalassia became brown and heavily fouled by algae for several months.

Another example of the oil spill effect on mangrove plants is December, 2000. In this case, 500 mangrove saplings in a 6.34 ha reserve in Hong Kong were subjected to a smuggled fuel oil spill. More than 80% died in root-zone sediments containing 60 - 80 ug/g total petroleum hydrocarbons (TPH). However, one year later, the injured survivors had recovered and re-grown, with root-zone sediment TPH concentrations approaching urban 'background' values of 40 ug/g TPH.

Impact of Oil Spills on Cyanobacteria and other Microorganisms

There are numerous factors that determine the microbial response to marine oil spills. These include the oil composition and degree of weathering, as well as the environmental conditions, particularly temperature and nutrient concentrations.

When crude oil is introduced into seawater, the microbial community changes and consists of multiple co-existing species. These can be explained by resource sharing. The diverse array of hydrocarbons present in crude oil requires resource partitioning by microbial populations, as well as microbial modification of oil components and the surrounding environment, will lead to temporal succession. The reviewers observed a network of direct and indirect interactions within and between species, even when just one type of hydrocarbon is present. They also provided a schematic illustration of some of the interactions involved in hydrocarbon biodegradation. Elements of these interactions were stated in several studies reviewed by them.

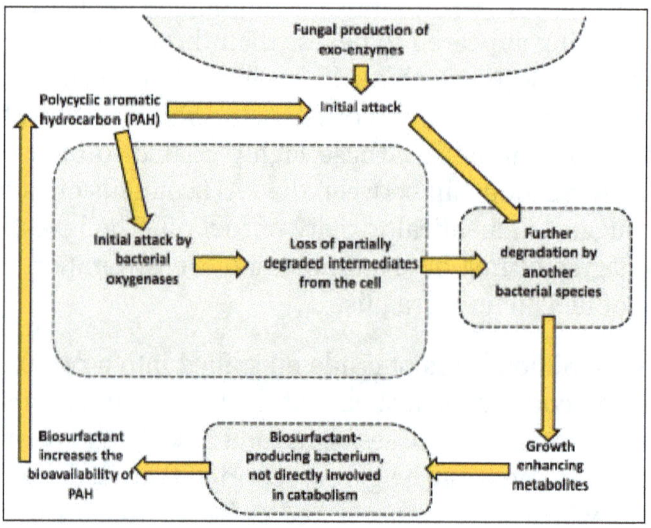

The interactions involved in hydrocarbon biodegradation.

Multiple Coexisting species and temporal succession are typically observed when Alcanivorax spp., which degrade straight-chain and branched alkane, are increased and found abundantly. This is followed by Cycloclasticusspp., which degrade PAHs. Other genera of obligate alkane degraders, such as Oleibacter sp. and Oceanospirillales sp., have been abundantly detected in other oil-rich marine environments.

At the molecular level, studies by the microarray analysis of 16S rRNA genes have revealed that Gamma-proteobacteria were enriched in marine environments that are historically exposed to frequent, episodic, natural 'oil spills'. This analysis also revealed that marine water exposed to oil spills enriches indigenous oil degrading bacteria, where members of the order Oceanospirillales comprised more than 90% of the bacterial community, compared to 5% of the uncontaminated sample. Experiments on the incubation and mixing of marine sediment with phenanthrene and bromodeoxyuridine (BDU), followed by the analysis of BDU-labelled DNA, revealed diverse groups of PAH degraders belonging to the genera Shewanella, Exiguobacterium, Methylomonas, Pseudomonas and Bacteroides. It also revealed Gammaproteobacteria and Deltaproteobacteria, which were not closely related to the cultivated organisms.

Following the 1991 Gulf War environmental disaster, an extensive formation of cyanobacterial mats was observed to colonize most of the oil polluted shores. Although most of the intertidal cyanobacterial mats were severely affected by the oil spill, these organisms were the first to recolonize the destroyed habitats. This initial massive growth of cyanobacteria, especially on sites where they did not occur before the oil spill, indicated the preference of the cyanobacterial mats to the absence of bioturbation (i.e., destabilizing the sediment surface caused by crabs and polychaetes) of the sediment for their growth. Cyanobacteria do not usually occur where bioturbation has been carried out and - together with the grazing pressure by benthic animals - prevents the establishment of cyanobacterial mats. The extensive growth of cyanobacteria following the oil pollution of the shores can be explained by destroying most of the crab colonies in the mudflats and immediately stopping the bioturbation process, as well as grazing by gastropods. Reference explained this colonization by three different processes. The first one is the desiccation, cracking and peeling of the cyanobacterial mats. This removes the uppermost part of the oiled sediment. The second is the resettlement of burrowing macrofauna like crabs and benthic animals, such as gastropods, which outcompete the cyanobacteria again. The third is further extensive growth of cyanobacteria building thick laminated mats. These layers completely seal the surface and hence, produce an anaerobic zone which inhibits oil degradation. As long as such a bloom of cyanobacteria exists, microbial oil degradation will be prevented. They will also prevent any resettlement by macrofauna.

Pollutants in Aquatic Invertebrates

Invertebrates constitute more than ninety-five per cent of the existing biodiversity of the world. They present an extreme range of variation in terms of their body architecture, adaptation, food preference, habitat preference, behavior and physiological response against environmental stressors. Evolutionarily, they are an ancient group of animals which have survived an extreme range of environmental adversities and perturbations from the time of their origin. Invertebrates occupied diverse kinds of habitats including terrestrial, freshwater, marine and estuarine ecosystems. Aquatic invertebrates are usually distributed in the multiple spatial compartments of the water bodies from the surface to the bottom region called the benthic stratum. Aquatic invertebrates are also found in the conspicuous ecological areas of the mud–water interface and subsoil region at the floor of aquifers. Freshwater ponds serve as habitat for diverse aquatic organisms including members of Porifera, Mollusca and Arthropoda. Majority of the ecotoxicological research was carried out on *Eunapius carteri.*

Aquatic invertebrates bear ecological, economical, nutritional and biotechnological significance and demand a special scientific attention. Many of them are the dietary source of nutrition for human and other organisms. Their importance as potential source of bioactive substances and pharmacologically active compounds cannot be denied. Current scientific reports indicate that the rapid and unrestricted contamination of freshwater and marine ecosystem by diverse pollutants poses serious ecotoxicological threat for the existence of invertebrates in their natural habitats. Various anthropogenic activities like habitat destruction and habitat contamination have been identified as the causative factors of dwindling of invertebrate species on earth. Continuous addition of various toxic chemical compounds into the global environment is a major environmental challenge encountered by human and other organisms. Many of these environmental pollutants are of either industrial or agricultural origin and need a thorough toxicological screening in invertebrate models.

The overall evolutionary success and survival efficacy of invertebrates depended on several factors including the development of a highly advanced immunological system. Immunology deals with the typical strategies of biological defence against environmental toxins and pathogens. Components of immune system and their coherent functional attributes enabled the invertebrates to overcome the toxin-induced chemical stresses of the primitive and modern global environment. Many of the environmental pollutants of the current hydrosphere are less researched, with limited toxicological information. Invertebrates over a period of time are assumed to evolve novel and unique modes of immunological reactivities to defend against the toxic insults of environmental pollutants at the cellular and molecular levels. Aquatic invertebrates are relatively a neglected group of organisms with reference to their immunotoxicological status in a biounsafe environment. The principal immunotoxicological responsiveness of invertebrates centre around the reactivities of circulating immunocytes and selected effector organs like gill, digestive gland, intestine, labial palp, etc.

Immunotoxins and their Impact on Aquatic Invertebrates

Immunotoxins are chemical compounds that can modulate the immune-related parameters and can adversely affect the biological and physiological functioning of the immune system of organisms including human. The freshwater ecosystem of India is being contaminated by diverse groups of xenobiotics of known and unknown chemistry. Ray reported mineral acids, alkalis, detergents, metalloids and pesticides as major environmental contaminants of the freshwater ecosystem. Scientists reported the toxic effects of pyrethroids, cypermethrin and fenvalerate in aquatic molluscan invertebrates and claimed them as immunotoxins for their adverse effect on the immune parameters of host animal. Toxicity of these pyrethroid pesticides on non-target aquatic invertebrates appears to be detrimental and can seriously affect their survival efficacy and reproductive success. Arsenic, another metalloid and a potent immunotoxin, bears toxic effects on aquatic invertebrates and can modulate the histopathology and immune parameters of freshwater mollusc and estuarine mud crab. Biopesticide azadirachtin has been recently identified as an immunotoxin due to its potential to affect the immunofunctioning of mussels. Moreover, during monsoon and flood, pesticide-laden agricultural runoff often contaminates the freshwater ecosystem and poses serious threat to its inhabitants. Washing soda, chemically known as anhydrous sodium carbonate, is an important aquatoxin that can alter the selected physiological parameters of diverse groups of invertebrates. Mukherjee reported the toxicity of washing soda in a freshwater sponge of India with reference to its phagocytic and cytotoxic status. Altered functioning of the immune system may lead to opportunistic invasion of environmental pathogens and parasites into the body of host and increase the vulnerability of these biofilter species in polluted environment.

The nature and magnitude of immunotoxicity depend on multiple parameters including the concentration of toxin, type and span of exposure and route of entry to the host. Immunotoxins are difficult to identify as they can cause a wide magnitude of adversity on the immune status of organisms. Immunotoxicology deals with the assessment of toxicological response in an organism by estimating the responsiveness and reactivity of its immunological parameters. Immunotoxicology of invertebrates, in recent times, has been gaining a special scientific attention for its efficacy in monitoring the health of environment – both aquatic and terrestrial. Accuracy and precision of selected toxicological responses of aquatic invertebrates enabled a few species to function as suitable biomonitoring agents of water pollution. Several effective immunomarkers of aquatic pollution have recently been established in model invertebrates.

External Physicochemical Barriers

External physicochemical barriers of the invertebrates act as the first line of defence against the invasion of parasite, pathogen and toxin in a polluted aquatic environment. External physicochemical barriers of the invertebrates include shell, tunic, test, cuticle, carapace, pinacoderm and others. In mollusc, hard calcareous shell or valves provide the primary protection against entry of pathogens and toxins. However, in shell-less molluscs, thick external cuticular sheet known as mantle or pallium are assumed to play a significant physiological role against toxin and pathogen entry. Soft body wall of molluscs consists of cuticle, epidermis and muscles, which are thought to take an active part in the innate immune defence against environmental pathogens and toxins. In a polluted environment, where environmental pathogens and toxins are present in sufficient quantity, the two valves of the mussel are kept closed to minimize the entry of undesired agents. In addition to external shell, mucus secreted from the internal viscera provides another line of defence against the invasion of pathogens and toxins. Mucus acts as a protective barrier preventing the direct contact of toxins to epithelia. Secretion of mucus is reported to be an important detoxification and evasive mechanism of invertebrates. Pathogens are trapped within the mucus secreted by the organism leading to elimination. Calcareous shell made up of calcium carbonate and mucus is considered as external physiochemical barrier and plays an important role in the immune defence of aquatic invertebrates. Carapace, the external physiochemical barrier of crab provides the first line of defence in crustaceans.

Corrosive toxins including mineral acids, alkalis, pesticides and detergents appear to be the potential threats of the invertebrate. Contamination of water bodies by these pollutants often results in breaches of the physicochemical barriers of external body surface, which leads to facilitation of invasion of toxic microorganisms and parasites into the viscera of target invertebrates. Components of innate immune response comprise physicochemical barrier against external pathogen and parasite entry. Cuticle covered with waxy material serves as a mechanical barrier of crab against parasitic infection. This chitinous exoskeleton or cuticle helps in the process of wound healing by preventing the fatal loss of haemolymph from body, maintenance of tissue architecture and prevention of opportunistic invasion of pathogen. Whenever this cuticle is damaged by injury or infection, the wound is rapidly sealed by clotting of immunocytes, preventing blood loss and pollutant entry. Once the clot is formed, wound is darkened and accumulation of melanin occurs. Melanin is reported to be involved in sealing of wound and synthesis of new cuticle. On the other hand, pinacoderm made up of flattened pinacocytes forms a continuous layer on the external surface of freshwater sponges and acts as a first line of defence against foreign invaders.

An experiment was performed to investigate the possible role of mucus as a primary barrier in removal of external particles. In this context, the shell of freshwater gastropod, *P. globosa* was drilled to create a micropore and a solution of cultured yeast (*Saccharomyces cerevisiae*) suspended in sterile phosphate buffered saline was injected into the body through a syringe. The mollusc was under scientific surveillance and the activity of the mollusc was photodocumented. After two hours of injection, hypersecretion of mucus was recorded by the experimental *P. globosa* within a time span of approximately thirty minutes. The expelled mucus contained more than ninety-five per cent of yeast particles, indicating the efficacy of mucus to act as external physicochemical barrier.

Xenobiotic-induced Shift in Haemocyte Density and Morphological Damages of Blood Cells.

Haemocytes are the immunocompetent cells which are functionally responsive to various xenobiotics present in the aquatic environment. Homeostasis of total haemocyte density of aquatic invertebrates within the permissible physiological limit may be considered as an important immunological parameter of cell-mediated immune response of molluscs. Chakraborty reported suppression in the total count of haemocytes of the freshwater edible mollusc, *L. marginalis* under the sublethal exposure of sodium arsenite, an aquatic pollutant. Mukherjee reported modulation in the total cell density of the same specimen under the sublethal concentrations of azadirachtin, a neem-based pesticide, a common contaminant of pond water. According to them, toxin-induced alteration in the total cell dynamics may lead to a gradual decline of this species in its natural habitat. The total count of haemocyte of *Villorita cyprinoides* was found to decrease under the exposure of copper. Ray reported dynamics of the total haemocyte density of *B. bengalensis* and *L. marginalis* exposed to environmentally realistic sublethal concentrations of cypermethrin and fenvalerate, respectively.

The differential cell count, on the other hand, is also considered as an important immunotoxicological marker of environmental pollution. Moreover, heterogenous populations of invertebrate blood cells have been reported to perform diverse physiological activities including nonself recognition, encapsulation and generation of cytotoxic molecules. Chakraborty reported that haemocyte density can be considered as a potential biomarker of toxicity of sodium arsenite, a vital pollutant of the Sunderbans estuary of India. Their experimental findings revealed the variation in the relative densities of granulocytes, agranulocytes, blast-like cells, hyalinocytes and asterocytes under the exposure of sodium arsenite. Ray reported a non-linear fluctuation in the differential cell density of *B. bengalensis* and *L. marginalis* following sublethal exposure of cypermethrin and fenvalerate, respectively. According to them, aquatic pesticide-induced alteration in the differential haemocyte density was indicative to stress-induced loss of cell balance, which might alter the physiological homeostasis of these species distributed in pesticide-contaminated habitat. Qubella proposed that fluctuation in the differential haemocyte density may be a result of reversible migration of haemocytes from tissues to haemolymph or *vice versa*.

Mukherjee reported an increase in relative percentage of pro-haemocytes subpopulation in aquatic mollusc *L. marginalis* under the sublethal exposure of azadirachtin, a contaminant of water bodies. According to them, pesticide-induced depletion in the densities of granulocytes, agranulocytes, hyalinocytes and asterocytes was indicative to a possible impairment of haemocytic function of mussels distributed in polluted environment. Ray flow cytometrically identified five discrete subpopulations of haemocytes, that is, blast-like cells, round hyalinocytes, spindle hyalinocytes, round granulocytes, and granular asterocytes in the freshwater molluscs, *B. bengalensis* and *L. marginalis*. Experimental exposure of 0.5 ppm of cypermethrin for 96 hours resulted in significant increase in the per cent populations of blast-like cells and round hyalinocytes and significant decrease in the subpopulation of granular asterocyte in *B. bengalensis* as compared to control. Exposure of 3 ppm of fenvalerate for 96 hours resulted in significant increase in the relative percentage of round granulocyte and significant decrease in the subpopulation of round hyalinocyte in *L. marginalis* as compared to control. Das reported alteration of relative densities of haemocyte subpopulations of *L. marginalis* under sublethal exposures of cypermethrin. Exposure to sublethal concentrations of cypermethrin yielded decrease in densities of agranulocytes and asterocytes and increase in densities of blast-like cells and granulocytes. Pyrethroids-induced alterations in the differential cell density of freshwater molluscs may lead to impairment of blood cell homeostasis in the experimental species distributed in polluted environment.

Morphological alterations of haemocytes under the exposure of environmental toxins play an important role in cell-mediated immune response of molluscs. Ray identified cytoplasmic hypervacuolation, rounding up of cell, alteration in cell shape, hypergranulation, increased cytoplasmic spreading, membrane disintegration and membrane blebbing as principal aberrations in the haemocytes of *B. bengalensis* and *L. marginalis*under the exposure of pollutant like pyrethroid. Cypermethrin- and fenvalerate-induced morphological damages in the circulating haemocytes of freshwater edible molluscs may lead to possible impairment of the functioning of cells and their immunological reactivity. Chakraborty and Ray reported dose- and time-dependent increase in the frequency of binucleated and micronucleated haemocytes and gill cells in the freshwater bivalve, *L. marginalis*, under the exposure of sodium arsenite.

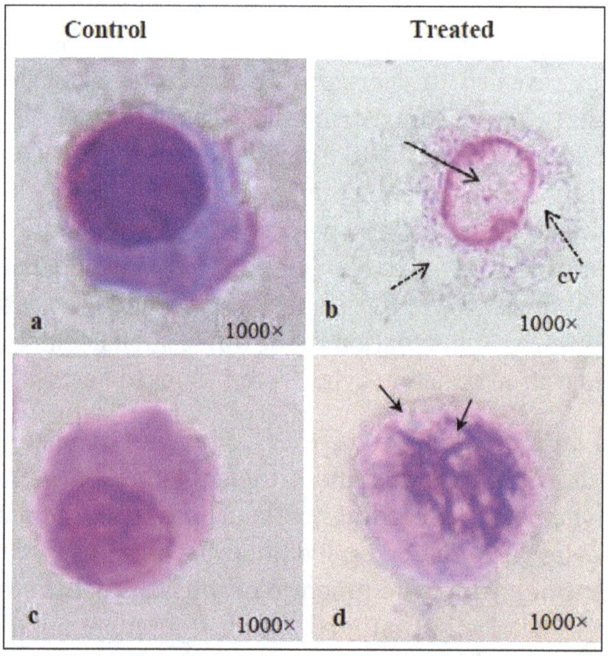

Photomicrographs of normal haemocytes of aquatic molluscs, P. globosa (a) and L. marginalis (c). Cypermethrin treatment (1.5 ppm/7 days) yielded intense cytoplasmic vacuolation (cv) (b) in the haemocytes of P. globosa and nuclear disintegration (d) in the haemocytes of L. marginalis.

Nonself Surface Adhesion and Aggregation Response of Haemocytes

Nonself surface adhesion is considered as an important immunological mechanism which is fundamentally related to self–nonself discrimination. Cell adhesion and interaction between cell and the substratum play a pivotal role in the development, maintenance and immune recognition in multicellular animals. Mukherjee studied the glass surface adhesion efficacy of haemocytes of *L. marginalis* under the sublethal concentrations of azadirachtin. According to them, the data were indicative to impairment of immunological response of *L. marginalis*in its natural habitat leading to a possible dwindling of this biofilter species from aquatic ecosystem. Ghosh studied the kinetics of nonself surface adhesion in the haemocytes of mud whelk, *Telescopium* exposed to diesel in the Sunderbans biosphere reserve. Vehicle diesel is reported to be a serious pollutant of estuary. Saha reported modulation in the nonself surface adhesion efficacy of haemocytes of edible mud crab, *Scylla serrata*, under the sublethal exposures of sodium arsenite. Ray reported arsenic-induced glass surface adhesion of haemocytes of juvenile mud crab. According to them, arsenic-induced alteration in the reactivity of haemocyte may render the juvenile mud crab to become immunologically impaired in the parasite- and pathogen-contaminated natural habitat.

In the dynamic freshwater ecosystem, the inhabitants often compete for niche for their better survival and propagation. Overlapping in the niche leads to a state of acute predation and fighting among animals. As a result, the animals may encounter acute competition and subsequent physical damage and loss of body fluid. Cellular aggregation is a functional attribute offered by the haemocytes of invertebrates to prevent the accidental blood loss by formation of biological plug at the site of wound and resist the entry of pathogenic microorganism. Selected pollutants like sodium arsenite, washing soda, pyrethroid, azadirachtin are reported to affect the aggregation response of many aquatic invertebrates. Hence, cell–cell aggregation is considered as an immunological response for host defence. Aggregation of haemocytes around invaded microorganisms is termed as "encapsulation response" and is considered as an important immunological response. When successful encapsulation occurs, a host animal can restrict the proliferative and invasive property of a pathogen. Encapsulation reaction is mediated by specific population of immunoactive blood cells and is effective in cell-mediated immunity of invertebrates. Mukherjee reported anticoagulant and carbohydrate-induced interference of cellular aggregation of mussel, *L. marginalis*, under experimental exposure of azadirachtin. Persistent exposure of azadirachtin inhibited the cellular aggregation response in freshwater mussel. Workers apprehend such scenario in the natural environment may lead to a decline in the population of freshwater mussel and loss of freshwater biodiversity of India. Furthermore, a drastic increase in the occurrence of free cells was recorded against ethylene diamine tetraacetic acid and manure treatment, which was suggestive to possible role of these chemical agents as inhibitor of cellular aggregation. Ray reported sodium-arsenite-induced inhibition in aggregation of haemocytes of juvenile mud crab, *S. serrata*. According to them, arsenic-induced alteration of immune status may impart a state of vulnerability in juvenile crab inhabiting the arsenic-polluted environment.

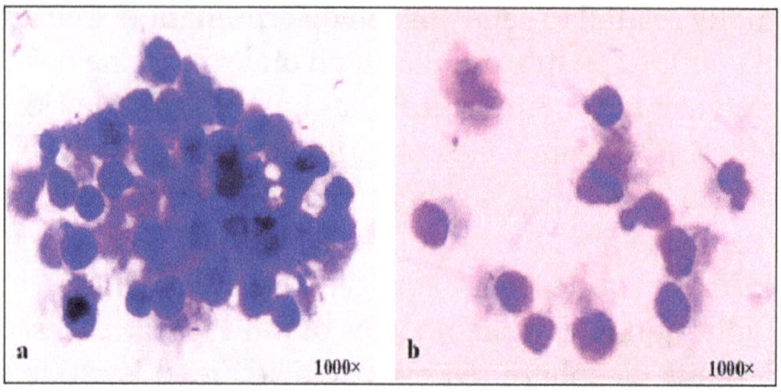

Typical aggregation response of haemocytes of a freshwater gastropod, Pila globosa (a). Cypermethrin is a contaminant of the habitat of P. globosa which inhibits the degree of haemocyte aggregation (b).

Phagocytic Response in the Face of Environmental Stressors

Phagocytosis, in general, is considered a classical innate immune response reported in the majority of the invertebrate Phyla. It is an established immunological response and is considered as a biomarker of aquatic pollution. Phagocytic response enables invertebrates to combat against invading pathogens and pollutants of known and unknown chemistry. Haemocyte-mediated phagocytosis of nonself particles provides natural immunity in the bivalves. Chakraborty reported the inhibitory effect of sodium arsenite on the phagocytic response of *L. marginalis* under the challenge of yeast at various sublethal concentrations. Inhibition in phagocytic response was also recorded under similar laboratory condition for the haemocytes of arsenic-treated *L. marginalis* when challenged with human red blood corpuscles.

According to them, impairment in the phagocytic potential of the haemocytes of arsenic-treated mussels may lead to compromisation of the immune status of the animals distributed in the contaminated habitat. Mukherjee reported azadirachtin-induced suppression in the phagocytic potential of haemocytes of *L. marginalis* under the experimental challenge of charcoal particulate. Azadirachtin-induced inhibition in the phagocytic response indicated a state of immune suppression of mussels. Ray screened the immunotoxicological reactivity of haemocytes of juvenile mud crab, *S. serrata* of the Sunderbans biosphere reserve and reported sodium-arsenite-induced inhibition in the phagocytic potential under the challenge of yeast. Arsenic-induced altered reactivity of haemocytes may affect the propagation and survival of mud crab population by increasing its vulnerability to higher rate of disease and parasite attack. Sponges, on the other hand, are non-selective filter feeders which depend on phagocytosis for the purposes of feeding and digestion. Phagocytosis, in sponge, plays a dual physiological attribute in the form of food procurement and innate immune response. Mukherjee reported washing-soda-induced inhibition in the phagocytic potential of cells of freshwater sponge, *E. carteri*, under the challenge of yeast. According to them, decrease in phagocytic potential is suggestive to a possible impairment of both food capture efficiency and innate immune status, leading to suppression of immunological status of *E. carteri* distributed in detergent-contaminated natural habitat.

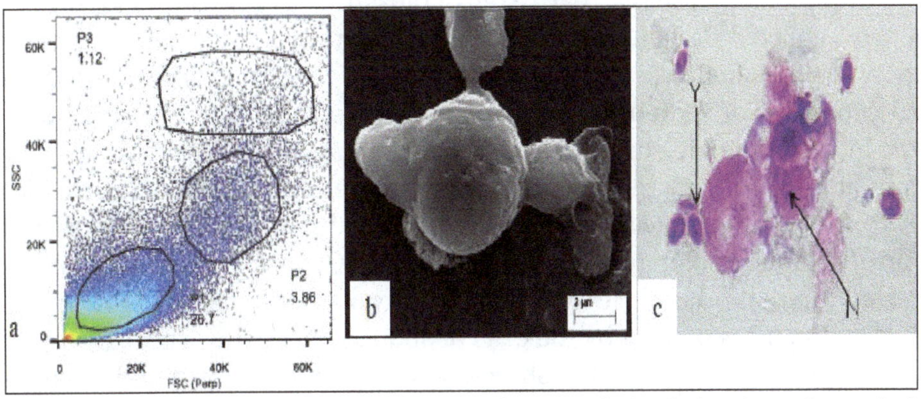

Flow cytometry of isolated cells of sponge, E. carteri, depicting their relative size and granularity (a). Phagocytosis of yeasts (Y) by the cells of E. carteri (b, c) exposed to washing soda. N = Nucleus.

Cytotoxicity of Blood Cells as an Effective Immune Strategy

Cytotoxic molecules generated by the immunocompetent cells of invertebrates are reported to play an important role in the destruction and deactivation of foreign engulfed pathogens. Contamination of natural habitat by toxic metals, metalloids, pesticides and washing soda results in a significant alteration in the cytotoxic status of invertebrates. Cytotoxic molecules are considered as an effective component of innate immune defence of invertebrates distributed in polluted environment. Nappi and Ottaviani reported nitric oxide and superoxide anions as potential "killing agents" of invertebrates. Generation of reactive oxygen intermediate and reactive nitrogen intermediate by the phagocytic cells are mediated by NADPH oxidase and nitric oxide synthase (NOS), respectively.

According to them, generation of superoxide anions can be correlated with increased respiratory burst activity in phagocytic cells. Mukherjee reported a dose-dependent increase in the generation of superoxide anion in the cells of freshwater edible bivalve, *L. marginalis* under the exposure of azadirachtin, a neem-based pesticide. According to them, this response of haemocytes may be considered as cellular stress under the sublethal and environmentally realistic concentrations of azadirachtin. Nitric oxide is considered as a signalling molecule generated in the biological system by the immunocytes of invertebrates as a potent cytotoxic agent for killing of invading microorganisms. It is generated during the conversion of L-arginine to L-citrulline by nitric oxide synthase. Saha reported sodium-arsenite-induced increase in the generation of the intracellular nitric oxide in estuarine mud crab, *S. serrata*. According to them, dose-dependent increase in the generation of nitric oxide may be considered as a biomarker of arsenic pollution. Chakraborty reported inhibition in generation of nitric oxide in *L. marginalis* under prolonged exposure of arsenic. Researchers claimed this studied parameter as possible biomarker of arsenic toxicity in aquatic environment. Phenoloxidase, on the other hand, is considered as another cytotoxic molecule and is involved in the process of nonself recognition, phagocytosis and melanisation. The production of toxic quinoid derivative by phenoloxidase is an early step of biosynthesis of melanin for host defence. Chakraborty reported depletion in generation of cytotoxic molecule like nitric oxide and activity of phenoloxidase in the gill of freshwater bivalve, *L. marginalis*, under the sublethal exposures of sodium arsenite. According to them, sodium-arsenite-induced alteration in the cytotoxic status in gill indicated a state of immunocompromisation in the animal. Chakraborty reported inhibition in the generation of superoxide anions and nitric oxide and activity of phenoloxidase in the digestive gland of the same experimental bivalve under prolonged exposure of arsenic. Generation of cytotoxic molecules like superoxide anions and nitric oxide and activity of phenoloxidase are important immunological responses offered by the cells of sponge under the exposure of pathogens and toxins. Researchers reported washing-soda-induced alteration in the generation of superoxide anion, nitric oxide and activity of phenoloxidase in the dissociated cells of freshwater sponge, *E. carteri*.Washing-soda-induced alteration in cytotoxic response of *E. carteri* may lead to an undesirable shift in the immune status of the animal distributed in detergent contaminated natural habitat.

Lysosomal Membrane Stability and Activity of Phosphatases

Lysosome is an important subcellular organelle involved in the process of degradation of foreign engulfed particulate. After phagocytosis, the phagocytic vacuole with engulfed foreign particles fuses with lysosome to form phagolysosome. As lysosome plays an important role in secretion of various digesting enzymes, maintenance of lysosomal membrane integrity has been gaining special scientific attention from the immunological point of view. Chakraborty and Ray reported impairment of lysosomal membrane integrity by neutral red retention assay in the haemocytes of *L. marginalis*exposed to sodium arsenite. According to them, arsenic-induced fragility of lysosomal membrane may lead to leakage of lysosomal enzymes into the cytosol and subsequent destruction of the adjoining self cell or tissue. Ray reported lysosomal membrane integrity of haemocytes of *B. bengalensis* and *L. marginalis* under the experimental exposures of cypermethrin and fenvalerate, respectively.

Researchers reported pyrethroid-induced lysosomal membrane fragility of haemocytes suggesting substantial damage and destabilization of lysosomal membrane. Screening of lysosomal membrane integrity of molluscan haemocytes by neutral red retention assay is considered as a possible biological marking of arsenic toxicity and as an early warning tool of environmental pollution. Phosphatases are principal lysosomal enzymes which are involved in pathogen destruction and are reported as markers of environmental stress. Chakraborty reported suppression in the activity of hydrolytic enzymes, acid and alkaline phosphatases in the gill of freshwater edible bivalve, *L. marginalis* under the exposure of sodium arsenite. According to them, inhibition in the activity of these enzymes might cripple the immune status and nutrient mobility in the gill of *L. marginalis*. Chakraborty reported inhibition in the activity of phosphatases in the haemocyte and digestive tissue of edible bivalve under similar toxic insult by environmental arsenic. Workers apprehend a state of immunocompromisation of the organisms under persistent exposure of arsenic. Saha reported inhibition in the activity of phosphatases in the haemocytes of estuarine mud crab, *S. serrata* under the exposure of arsenic. According to them, this situation in the natural environment might result in impairment of immunological activity and opportunistic invasion of parasite and pathogen into the body of the organism.

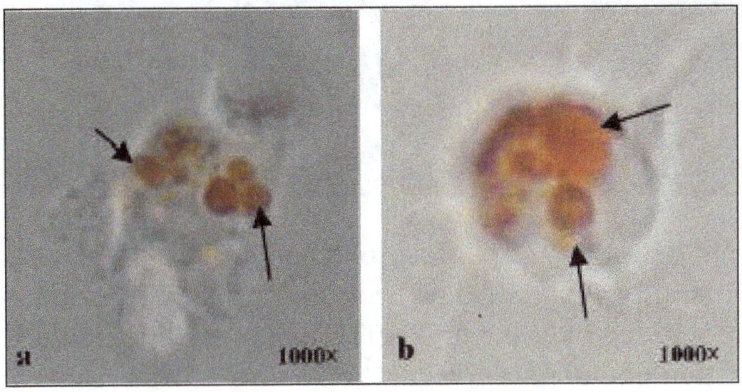

Cypermethrin-induced fragility of lysosome membrane of haemocytes of aquatic mollusc as evident from neutral red dye retention assay. Dye neutral red concentrated in the lysosomal compartments at zero minute of the assay (a). Complete diffusion of neutral red in the cytoplasm of cypermethrin-exposed haemocytes has been designated as "end point" of the assay (b).

Pollutant-induced Apoptosis and Necrosis of Invertebrate Immunocytes

Apoptosis, the programmed cell death, is functionally involved in developmental and immunological processes of organisms. Moreover, apoptosis helps in the process of removal of damaged cells and thus is considered an important machinery of survival of host animal in polluted environment. It is involved in specific cell signaling mechanism, which is yet to be studied in invertebrates in detail. However, pesticides and other environmental xenobiotics are reported to affect the apoptotic and necrotic pathways of the immunoactive cells of invertebrates. Current scientific reports suggest apoptosis as an effective biomarker of aquatic pollution. Apoptosis presented changes in the morphological characteristics of cells including membrane blebbing, nuclear condensation, cytoplasmic shrinkage and membrane asymmetry. Kiss reported the translocation of phosphatidylserine from inner leaflet of the plasma membrane to the outer leaftet, which was considered a hallmark of apoptosis. Ray reported the apoptotic and necrotic cell deaths of haemocytes of *B. bengalensis*and *L. marginalis* under the sublethal exposures of cypermethrin and fenvalerate, respectively, employing flow cytometry. According to them, cypermethrin and fenvalerate treatment yielded decrease in the percentage of apoptotic and necrotic haemocyte morphotypes of *B. bengalensis* and *L. marginalis*. Pyrethroid-induced apoptosis of molluscan haemocytes is considered as impairment of immunological status of *B. bengalensis* and *L. marginalis*.

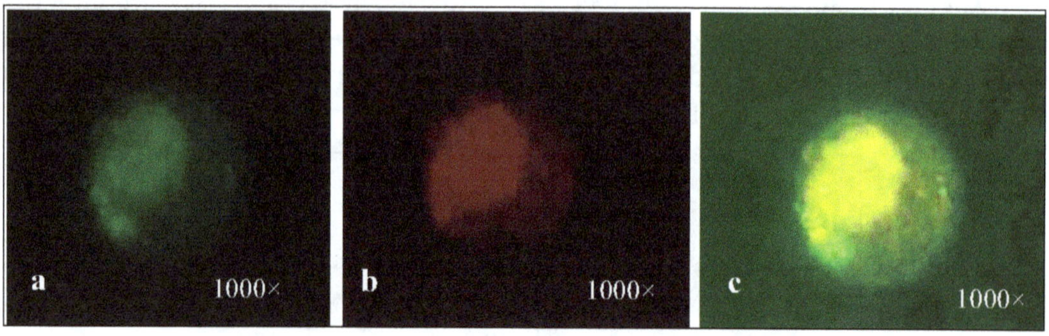

Immunofluorescent detection of apoptosis of haemocytes of pond snail, B. bengalensis exposed to sublethal (3000 ppm/7 days) concentrations of washing soda. (a) and (b) represent positive staining by Annexin V-FITC and propidium iodide, respectively; (c) depicts co-localization of fluorescent signals.

Toxin-induced Damage of Target Organs and Tissues

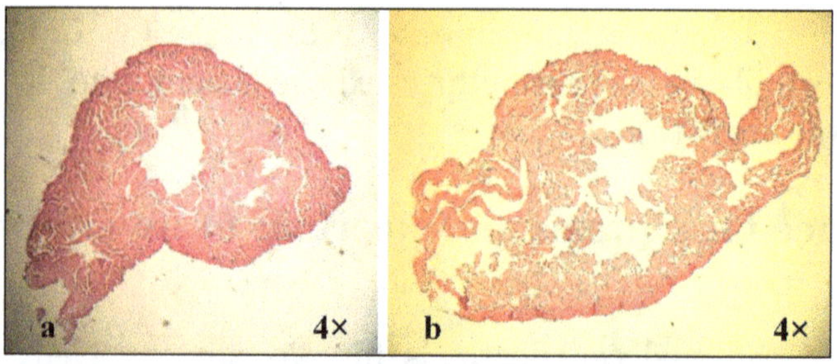

Cypermethrin (1.5 ppm/7 days)-induced histopathological damage of heart of mollusc, P. globosa.

For filter feeding aquatic organisms, the gill, heart, digestive gland, mantle and antennae act as major target organs of common environmental xenobiotics. Mollusca bear a well-developed gill which serves as an organ of gaseous exchange, filter feeding and immunosurveillance. Molluscan gill is a thin membranous and vascularized organ that is in continuous contact with the environmental toxicants distributed in water. During the process of respiration and filter feeding, gill is exposed to various toxins of known and unknown chemistry. Exposure of gill to aquatic sodium arsenite, pyrethroid pesticide and washing soda of pond water yields a structural damage of lamellae. Chakraborty reported hyperchromatic anaplastic cells in the gills of *L. marginalis* exposed to sublethal concentrations of arsenic. This is often associated with tissue rupture and formation of dense fibrosis. The heart, in general, acts as an organ of body fluid pumping.

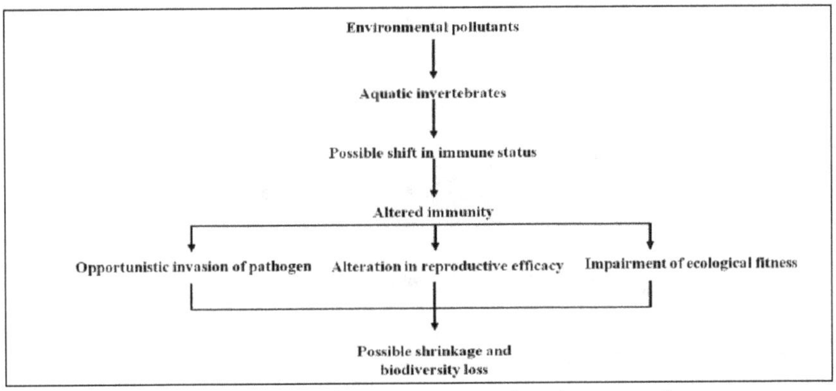

Ecophysiological consequences of environmental immunotoxins on aquatic invertebrates.

Exposure of cypermethrin results in severe histopathological damage of heart of mollusc. Similar kind of morphological damage was recorded due to exposure of aquatic mollusc to washing soda, a common pollutant of pond water. Digestive gland of mollusc bears immense physiological significance. The organ acts as an important site of diverse metabolic activities and biochemical detoxification. Exposure of sodium arsenite yields histopathological damage of the digestive gland of *L. marginalis*. Toxin-induced morphological alteration of digestive gland of mollusc includes hyperinfiltration of haemocytes, vacuolation and inflammatory lysis.

References

- Aquatic-toxicology, pharmacology-toxicology-and-pharmaceutical-science: sciencedirect.com, Retrieved 15 August, 2020
- Impact-of-oil-spills-on-marine-life, emerging-pollutants-in-the-environment-current-and-further-implications: intechopen.com, Retrieved 05 January, 2020
- Feely, R. A.; Sabine, C. L.; Hernandez-Ayon, J. M.; Ianson, D.; Hales, B. (June 2008). "Evidence for upwelling of corrosive "acidified" water onto the continental shelf". Science. 320 (5882): 1490–2. Bibcode:2008Sci...320.1490F. CiteSeerX 10.1.1.328.3181. doi:10.1126/science.1155676. PMID 18497259. Retrieved 25 January 2014
- Immunotoxicological-threats-of-pollutants-in-aquatic-invertebrates, emerging-pollutants-in-the-environment-current-and-further-implications: intechopen.com, Retrieved 13 May, 2020

5
Monitoring, Testing and Treatment

Various testing and treatment methods are used for the monitoring of environmental toxicity. A few of such methods are bioremediation, phyto-remediation, environmental sampling, etc. All these monitoring, testing and treatment methods have been carefully analyzed in this chapter.

Environmental monitoring is needed to be able protect the public and the environment from toxic contaminants and pathogens. It is used to prepare environmental impact assessments, as well establish the circumstances in which human activities carry a risk of harmful effects on the natural environment.

Types of monitoring carried out includes that of air, water and soil. All monitoring strategies and programmes have reasons and justifications which help them to establish the current status of an environment or to establish trends in environmental parameters.

Types of Contamination

Toxic contaminants and pathogens that can be released into a variety of media including air, soil, and water.

Air pollutants originate from sources such as vehicle emissions, refineries, power plants and industrial and laboratory processes. Pollutants include sulfur dioxide, nitrogen dioxide, carbon monoxide and volatile organic compounds.

Soil and water contaminants can be split into microbiological, radioactive, inorganic, synthetic organic and volatile organic compounds. Pesticides and herbicides are applied directly to plants and soils, and spills, leaking pipes, underground storage tanks, waste dumps, and waste repositories can cause incidental releases of other contaminants.

Heavy metals and organic contaminants are increasing in parts of the environment as a result of anthropogenic activity. Metal contaminants concentrate in aquatic matrices such as suspended matter, sediment and biota, which results in an increased presence in the aquatic food chain.

Some contaminants persist for years and can migrate through large regions of soil, until they eventually reach water resources and present ecological and human-health threats.

Types of Monitoring

The three main types of environmental monitoring are: Air, Soil and Water.

Air pollution is a constantly growing problem worldwide. Air monitoring is done via air sampling. The sampling can be passive, diffusive or done by biomonitoring with organisms that bioaccumulate air pollutants. As air pollution concentration is heavily influenced by the wind, anemometer data is also taken into account, as well as topography as landscape features can hinder the process of lateral atmospheric mixing.

Soil monitoring is done through soil sampling, which includes looking at soil erosion, salinity and contamination. Farming key to the world's food production while the regeneration of forests and jungles is central to keeping the air clean and free of CO_2, so a large number of environmental monitoring projects focus on soil quality.

Water monitoring has a wide range of sampling methods depending on the environment and the type of material being analysed. The types of water sampling include judgmental, simple random, stratified, grid, systematic, adaptive cluster, biomonitoring, remote sensors, passive, semi and continuous and grab sampling. The chemical condition of water is of high importance, so the presence of oxygen, nutrients, oils, pesticides and metals are all monitored.

Types of Sensors

There are numerous types of environmental sensors depending on the specifications needed for monitoring. Sensitivity, size and speed are all factors to consider when choosing an environmental sensor. Sensors can be divided into categories depending on the type of sensor being used, and the environmental factors that an analyst is looking at.

The main types of sensors include:

- Trace metal sensors: Laser-induced breakdown spectroscopy (LIBS), nanoelectrode array and miniature chemical flow-probe sensor.
- Radioisotopic sensors: RadFET (Radiation field-effect transistor), cadmium zinc telluride (CZT) detectors, low-energy pin diodes beta spectrometer, thermoluminescent dosimeter (TLD), isotope identification gamma detector, neutron generator for nuclear material detection, non-sandia radiation detectors.
- Volatile organic compound sensors: Evanescent fiber-optic chemical sensor, grating light reflection spectroscopy (GLRS), miniature chemical flow probe sensor, chemical sensor arrays, MicroChemLab (gas phase), gold nanoparticle chemiresistors, electrical impedance of tethered lipid bilayers on planar electrodes, MicroHound, hyperspectral imaging, chemiresistor array.

- Biological sensors: Fatty acid methyl esters (FAME) analyser, insulator-based dielectrophoresis, biological sensor arrays, μProLab, MicroChemLab (Liquid).

Sensors can also be split into types depending on environmental factors. These types of sensors include:

- Moisture sensors: Moisture sensors are essential for the measurement of volumetric water content in the soil towards the chimerical environment by several orders of magnitude.
- Pressure sensors/transducers: Pressure sensors operate on the basis of pressure applied. They are used in fields such as aviation, manufacturing biomedical measurements, auto mobile and hydraulic measurements.
- Tilt sensors: The tilt sensor relates the two different axes indicating a reference plane which may be in two different axes. This plays an important role in measuring tilt angles with reference to the Earth's ground plane. Tilt sensors are common in industry and in game controllers.
- Rain sensors: Rain sensors are a type of switching device which gets activated in the presence of rainfall. Rain sensors are used in areas such as as water conservation devices in irrigation systems, and in automobiles with windscreen wipers.
- Vibration sensors: Vibration sensors are important for displacement and acceleration with respect to the enduring impact on the environment.

Monitoring Environmental Contaminants

Air Quality Monitoring

Air pollutants are atmospheric substances—both naturally occurring and anthropogenic—which may potentially have a negative impact on the environment and organism health. With the evolution of new chemicals and industrial processes has come the introduction or elevation of pollutants in the atmosphere, as well as environmental research and regulations, increasing the demand for air quality monitoring.

Air quality monitoring is challenging to enact as it requires the effective integration of multiple environmental data sources, which often originate from different environmental networks and institutions. These challenges require specialized observation equipment and tools to establish air pollutant concentrations, including sensor networks, geographic information system (GIS) models, and the Sensor Observation Service (SOS), a web service for querying real-time sensor data. Air dispersion models that combine topographic, emissions, and meteorological data to predict air pollutant concentrations are often helpful in interpreting air monitoring data. Additionally, consideration of anemometer data in the area between sources and the monitor often provides insights on the source of the air contaminants recorded by an air pollution monitor.

Air quality monitoring station

Air quality monitors are operated by citizens, regulatory agencies, and researchers to investigate air quality and the effects of air pollution. Interpretation of ambient air monitoring data often involves a consideration of the spatial and temporal representativeness of the data gathered, and the health effects associated with exposure to the monitored levels. If the interpretation reveals concentrations of multiple chemical compounds, a unique "chemical fingerprint" of a particular air pollution source may emerge from analysis of the data.

Air Sampling

Passive or "diffusive" air sampling depends on meteorological conditions such as wind to diffuse air pollutants to a sorbent medium. Passive samplers have the advantage of typically being small, quiet, and easy to deploy, and they are particularly useful in air quality studies that determine key areas for future continuous monitoring.

Air pollution can also be assessed by biomonitoring with organisms that bioaccumulate air pollutants, such as lichens, mosses, fungi, and other biomass. One of the benefits of this type of sampling is how quantitative information can be obtained via measurements of accumulated compounds, representative of the environment from which they came. However, careful considerations must be made in choosing the particular organism, how it's dispersed, and relevance to the pollutant.

Other sampling methods include the use of a denuder, needle trap devices, and microextraction techniques.

Soil Monitoring

Soil monitoring involves the collection and analysis of soil and its associated quality, constituents, and physical status to determine or guarantee its fitness for use. Soil faces many threats, including compaction, contamination, organic material loss, biodiversity loss, slope stability issues, erosion, salinization, and acidification. Soil monitoring helps characterize these threats and other potential risks to the soil, surrounding environments, animal health, and human health.

Collecting a soil sample in Mexico for pathogen testing.

Assessing these threats and other risks to soil can be challenging due to a variety of factors, including soil's heterogeneity and complexity, scarcity of toxicity data, lack of understanding of a contaminant's fate, and variability in levels of soil screening. This requires a risk assessment approach and analysis techniques that prioritize environmental protection, risk reduction, and, if necessary, remediation methods. Soil monitoring plays a significant role in that risk assessment, not only aiding in the identification of at-risk and affected areas but also in the establishment of base background values of soil.

Soil monitoring has historically focused on more classical conditions and contaminants, including toxic elements (e.g., mercury, lead, and arsenic) and persistent organic pollutants (POPs). Historically, testing for these and other aspects of soil, however, has had its own set of challenges, as sampling in most cases is of a destructive in nature, requiring multiple samples over time. Additionally, procedural and analytical errors may be introduced due to variability among references and methods, particularly over time. However, as analytical techniques evolve and new knowledge about ecological processes and contaminant effects disseminate, the focus of monitoring will likely broaden over time and the quality of monitoring will continue to improve.

Soil Sampling

The two primary types of soil sampling are grab sampling and composite sampling. Grab sampling involves the collection of an individual sample at a specific time and place, while composite sampling involves the collection of a homogenized mixture of multiple individual samples at either a specific place over different times or multiple locations at a specific time. Soil sampling may occur both at shallow ground levels or deep in the ground, with collection methods varying by level collected from. Scoops, augers, core barrel, and solid-tube samplers, and other tools are used at shallow ground levels, whereas split-tube, solid-tube, or hydraulic methods may be used in deep ground.

Monitoring Programs

Soil Contamination Monitoring

A portable X-ray fluorescence (XRF) analyzer can be used in the field for testing soils for metal contamination.

Soil contamination monitoring helps researchers identify patterns and trends in contaminant deposition, movement, and effect. Human-based pressures such as tourism, industrial activity, urban sprawl, construction work, and inadequate agriculture/forestry practices can contribute to and make worse soil contamination and lead to the soil becoming unfit for its intended use. Both inorganic and organic pollutants may make their way to the soil, having a wide variety of detrimental effects. Soil contamination monitoring is therefore important to identify risk areas, set baselines, and identify contaminated zones for remediation. Monitoring efforts may range from local farms to nationwide efforts, such as those made by China in the late 2000s, providing details such as the nature of contaminants, their quantity, effects, concentration patterns, and remediation feasibility. Monitoring and analytical equipment will ideally will have high response times, high levels of resolution and automation, and a certain degree of self-sufficiency. Chemical techniques may be used to measure toxic elements and POPs using chromatography and spectrometry, geophysical techniques may assess physical properties of large terrains, and biological techniques may use specific organisms to gauge not only contaminant level but also byproducts of contaminant biodegradation. These techniques and others are increasingly becoming more efficient, and laboratory instrumentation is becoming more precise, resulting in more meaningful monitoring outcomes.

Soil Erosion Monitoring

Soil erosion monitoring helps researchers identify patterns and trends in soil and sediment movement. Monitoring programs have varied over the years, from long-term academic research on university plots to reconnaissance-based surveys of biogeoclimatic areas. In most methods, however, the general focus is on identifying and measuring all the dominant erosion processes in a given area. Additionally, soil erosion monitoring may attempt to quantify the effects of erosion on crop productivity, though challenging "because of the many complexities in the relationship between soils and plants and their management under a variable climate."

Soil Salinity Monitoring

Soil salinity monitoring helps researchers identify patterns and trends in soil salt content. Both the natural process of seawater intrusion and the human-induced processes of inappropriate soil and water management can lead to salinity problems in soil, with up to one billion hectares of land affected globally. Salinity monitoring at the local level may look closely at the root zone to gauge salinity impact and develop management options, whereas at the regional and national level salinity monitoring may help with identifying areas at-risk and aiding policymakers in tackling the issue before it spreads. The monitoring process itself may be performed using technologies such as remote sensing and geographic information systems (GIS) to identify salinity via greenness, brightness, and whiteness at the surface level. Direct analysis of soil up close, including the use of electromagnetic induction techniques, may also be used to monitor soil salinity.

Water Quality Monitoring

Design of Environmental Monitoring Programmes

Electrofishing survey methods use a mild electric shock to temporarily stun fish for capture, identification and counting. The fish are then returned to the water unharmed.

Water quality monitoring is of little use without a clear and unambiguous definition of the reasons for the monitoring and the objectives that it will satisfy. Almost all monitoring (except perhaps remote sensing) is in some part invasive of the environment under study and extensive and poorly planned monitoring carries a risk of damage to the environment. This may be a critical consideration in wilderness areas or when monitoring very rare organisms or those that are averse to human presence. Some monitoring techniques, such as gill netting fish to estimate populations, can be very damaging, at least to the local population and can also degrade public trust in scientists carrying out the monitoring.

Almost all mainstream environmentalism monitoring projects form part of an overall monitoring strategy or research field, and these field and strategies are themselves derived from the high levels objectives or aspirations of an organisation. Unless individual monitoring projects fit into a wider strategic framework, the results are unlikely to be published and the environmental understanding produced by the monitoring will be lost.

Parameters

Chemical

Analyzing water samples for pesticides: The range of chemical parameters that have the potential to affect any ecosystem is very large and in all monitoring programmes it is necessary to target a suite of parameters based on local knowledge and past practice for an initial review. The list can be expanded or reduced based on developing knowledge and the outcome of the initial surveys.

Freshwater environments have been extensively studied for many years and there is a robust understanding of the interactions between chemistry and the environment across much of the world. However, as new materials are developed and new pressures come to bear, revisions to monitoring programmes will be required. In the last 20 years acid rain, synthetic hormone analogues, halogenated hydrocarbons, greenhouse gases and many others have required changes to monitoring strategies.

Biological

In ecological monitoring, the monitoring strategy and effort is directed at the plants and animals in the environment under review and is specific to each individual study.

However, in more generalised environmental monitoring, many animals act as robust indicators of the quality of the environment that they are experiencing or have experienced in the recent past. One of the most familiar examples is the monitoring of numbers of Salmonid fish such as brown trout or Atlantic salmon in river systems and lakes to detect slow trends in adverse environmental effects. The steep decline in salmonid fish populations was one of the early indications of the problem that later became known as acid rain.

In recent years much more attention has been given to a more holistic approach in which the ecosystem health is assessed and used as the monitoring tool itself. It is this approach that underpins the monitoring protocols of the Water Framework Directive in the European Union.

Radiological

Radiation monitoring involves the measurement of radiation dose or radionuclide contamination for reasons related to the assessment or control of exposure to ionizing radiation or radioactive substances, and the interpretation of the results. The 'measurement' of dose often means the measurement of a dose equivalent quantity as a proxy (i.e. substitute) for a dose quantity that cannot be measured directly. Also, sampling may be involved as a preliminary step to measurement of the content of radionuclides in environmental media. The methodological and technical details of the design and operation of monitoring programmes and systems for different radionuclides, environmental media and types of facility are given in IAEA Safety Guide RS–G-1.8 and in IAEA Safety Report No. 64.

Radiation monitoring is often carried out using networks of fixed and deployable sensors such as the US Environmental Protection Agency's Radnet and the SPEEDI network in Japan. Airborne surveys are also made by organizations like the Nuclear Emergency Support Team.

Microbiological

Bacteria and viruses are the most commonly monitored groups of microbiological organisms and even these are only of great relevance where water in the aquatic environment is subsequently used as drinking water or where water contact recreation such as swimming or canoeing is practised.

Although pathogens are the primary focus of attention, the principal monitoring effort is almost always directed at much more common indicator species such as *Escherichia coli*, supplemented by overall coliform bacteria counts. The rationale behind this monitoring strategy is that most human pathogens originate from other humans via the sewage stream. Many sewage treatment plants have no sterilisation final stage and therefore discharge an effluent which, although having a clean appearance, still contains many millions of bacteria per litre, the majority of which are relatively harmless coliform bacteria. Counting the number of harmless (or less harmful) sewage bacteria allows a judgement to be made about the probability of significant numbers of pathogenic bacteria or viruses being present. Where *E. coli* or coliform levels exceed pre-set trigger values, more intensive monitoring including specific monitoring for pathogenic species is then initiated.

Populations

Monitoring strategies can produce misleading answers when relaying on counts of species or presence or absence of particular organisms if there is no regard to population size. Understanding the populations dynamics of an organism being monitored is critical.

As an example if presence or absence of a particular organism within a 10 km square is the measure adopted by a monitoring strategy, then a reduction of population from 10,000 per square to 10 per square will go unnoticed despite the very significant impact experienced by the organism.

Monitoring Programmes

All scientifically reliable environmental monitoring is performed in line with a published programme. The programme may include the overall objectives of the organisation, references to the specific strategies that helps deliver the objective and details of specific projects or tasks within those strategies the key feature of any programme is the listing of what is being monitored and how that monitoring is to take place and the time-scale over which it should all happen. Typically, and often as an appendix, a monitoring programme will provide a table of locations, dates and sampling methods that are proposed and which, if undertaken in full, will deliver the published monitoring programme.

There are a number of commercial software packages which can assist with the implementation of the programme, monitor its progress and flag up inconsistencies or omissions but none of these can provide the key building block which is the programme itself.

Environmental Monitoring Data Management Systems

Given the multiple types and increasing volumes and importance of monitoring data, commercial software Environmental Data Management Systems (EDMS) or E-MDMS are increasingly in common use by regulated industries. They provide a means of managing all monitoring data in a single central place. Quality validation, compliance checking, verifying all data has been received, and sending alerts are generally automated. Typical interrogation functionality enables comparison of data sets both temporarily and spatially. They will also generate regulatory and other reports.

One formal certification scheme exists specifically for environmental data management software. This is provided by the Environment Agency in the U.K. under its Monitoring Certification Scheme (MCERTS).

Sampling Methods

There are a wide range of sampling methods which depend on the type of environment, the material being sampled and the subsequent analysis of the sample.

At its simplest a sample can be filling a clean bottle with river water and submitting it for conventional chemical analysis. At the more complex end, sample data may be produced by complex electronic sensing devices taking sub-samples over fixed or variable time periods.

Judgmental Sampling

In judgmental sampling, the selection of sampling units (i.e., the number and location and timing of collecting samples) is based on knowledge of the feature or condition under investigation and on professional judgment. Judgmental sampling is distinguished from probability-based sampling in that inferences are based on professional judgment, not statistical scientific theory. Therefore, conclusions about the target population are limited and depend entirely on the validity and accuracy of professional judgment; probabilistic statements about parameters are not possible. Expert judgment may also be used in conjunction with other sampling designs to produce effective sampling for defensible decisions.

Simple Random Sampling

In simple random sampling, particular sampling units (for example, locations and times) are selected using random numbers, and all possible selections of a given number of units are equally likely. For example, a simple random sample of a set of drums can be taken by numbering all the drums and randomly selecting numbers from that list or by sampling an area by using pairs of random coordinates. This method is easy to understand, and the equations for determining sample size are relatively straightforward. This figure illustrates a possible simple random sample for a square area of soil. Simple random sampling is most useful when the population of interest is relatively homogeneous; i.e., no major patterns of contamination or "hot spots" are expected. The main advantages of this design are:

- It provides statistically unbiased estimates of the mean, proportions, and variability.
- It is easy to understand and easy to implement.
- Sample size calculations and data analysis are very straightforward.

In some cases, implementation of a simple random sample can be more difficult than some other types of designs (for example, grid samples) because of the difficulty of precisely identifying random geographic locations. Additionally, simple random sampling can be more costly than other plans if difficulties in obtaining samples due to location causes an expenditure of extra effort.

Stratified Sampling

In stratified sampling, the target population is separated into non-overlapping strata, or subpopulations that are known or thought to be more homogeneous (relative to the environmental medium or the contaminant), so that there tends to be less variation among sampling units in the same stratum than among sampling units in different strata. Strata may be chosen on the basis of spatial or temporal proximity of the units, or on the basis of preexisting information or professional judgment about the site or process. Advantages of this sampling design are that it has potential for achieving greater precision in estimates of the mean and variance, and that it allows computation of reliable estimates for population subgroups of special interest. Greater precision can be obtained if the measurement of interest is strongly correlated with the variable used to make the strata.

Systematic and Grid Sampling

In systematic and grid sampling, samples are taken at regularly spaced intervals over space or time. An initial location or time is chosen at random, and then the remaining sampling locations are defined so that all locations are at regular intervals over an area (grid) or time (systematic). Examples Systematic Grid Sampling - Square Grid Systematic Grid Sampling - Triangular Grids of systematic grids include square, rectangular, triangular, or radial grids. In random systematic sampling, an initial sampling location (or time) is chosen at random and the remaining sampling sites are specified so that they are located according to a regular pattern. Random systematic sampling is used to search for hot spots and to infer means, percentiles, or other parameters and is also useful for estimating spatial patterns or trends over time. This design provides a practical and easy method for designating sample locations and ensures uniform coverage of a site, unit, or process.

Ranked set sampling is an innovative design that can be highly useful and cost efficient in obtaining better estimates of mean concentration levels in soil and other environmental media by explicitly incorporating the professional judgment of a field investigator or a field screening measurement method to pick specific sampling locations in the field. Ranked set sampling uses a two-phase sampling design that identifies sets of field locations, utilizes inexpensive measurements to rank locations within each set, and then selects one location from each set for sampling. In ranked set sampling, m sets (each of size r) of field locations are identified using simple random sampling. The locations are ranked independently within each set using professional judgment or inexpensive, fast, or surrogate measurements. One sampling unit from each set is then selected (based on the observed ranks) for subsequent measurement using a more accurate and reliable (hence, more expensive) method for the contaminant of interest. Relative to simple random sampling, this design results in more representative samples and so leads to more precise estimates of the population parameters. Ranked set sampling is useful when the cost of locating and ranking locations in the field is low compared to laboratory measurements. It is also appropriate when an inexpensive auxiliary variable (based on expert knowledge or measurement) is available to rank population units with respect to the variable of interest. To use this design effectively, it is important that the ranking method and analytical method are strongly correlated.

Adaptive Cluster Sampling

In adaptive cluster sampling, samples are taken using simple random sampling, and additional samples are taken at locations where measurements exceed some threshold value. Several additional rounds of sampling and analysis may be needed. Adaptive cluster sampling tracks the selection probabilities for later phases of sampling so that an unbiased estimate of the population mean can be calculated despite oversampling of certain areas. An example application of adaptive cluster sampling is delineating the borders of a plume of contamination. Adaptive sampling is useful for estimating or searching for rare characteristics in a population and is appropriate for inexpensive, rapid measurements. It enables delineating the boundaries of hot spots, while also using all data collected with appropriate weighting to give unbiased estimates of the population mean.

Grab Samples

Collecting a grab sample on a stream.

Grab samples are samples taken of a homogeneous material, usually water, in a single vessel. Filling a clean bottle with river water is a very common example. Grab samples provide a good snap-shot view of the quality of the sampled environment at the point of sampling and at the time of sampling. Without additional monitoring, the results cannot be extrapolated to other times or to other parts of the river, lake or ground-water.

In order to enable grab samples or rivers to be treated as representative, repeat transverse and longitudinal transect surveys taken at different times of day and times of year are required to establish that the grab-sample location is as representative as is reasonably possible. For large rivers such surveys should also have regard to the depth of the sample and how to best manage the sampling locations at times of flood and drought.

A rosette sampler used for ocean monitoring.

In lakes grab samples are relatively simple to take using depth samplers which can be lowered to a pre-determined depth and then closed trapping a fixed volume of water from the required depth. In all but the shallowest lakes, there are major changes in the chemical composition of lake water at different depths, especially during the summer months when many lakes stratify into a warm, well oxygenated upper layer (*epilimnion*) and a cool de-oxygenated lower layer *(hypolimnion)*.

In the open seas marine environment grab samples can establish a wide range of baseline parameters such as salinity and a range of cation and anion concentrations. However, where changing conditions are an issue such as near river or sewage discharges, close to the effects of volcanism or close to areas of freshwater input from melting ice, a grab sample can only give a very partial answer when taken on its own.

Semi-continuous Monitoring and Continuous

An automated sampling station and data logger (to record temperature, specific conductance, and dissolved oxygen levels).

There is a wide range of specialized sampling equipment available that can be programmed to take samples at fixed or variable time intervals or in response to an external trigger. For example, a sampler can be programmed to start taking samples of a river at 8-minute intervals when the rainfall intensity rises above 1 mm/hour. The trigger in this case may be a remote rain gauge communicating with the sampler by using cell phone or meteor burst technology. Samplers can also take individual discrete samples at each sampling occasion or bulk up samples into composite so that in the course of one day, such a sampler might produce 12 composite samples each composed of 6 sub-samples taken at 20-minute intervals.

Continuous or quasi-continuous monitoring involves having an automated analytical facility close to the environment being monitored so that results can, if required, be viewed in real time. Such systems are often established to protect important water supplies such as in the River Dee regulation system but may also be part of an overall monitoring strategy on large strategic rivers where early warning of potential problems is essential. Such systems routinely provide data on parameters such as pH, dissolved oxygen, conductivity, turbidity and colour but it is also possible to operate gas liquid chromatography with mass spectrometry technologies (GLC/MS) to examine a wide range of potential organic pollutants. In all examples of automated bank-side analysis there is a requirement for water to be pumped from the river into the monitoring station. Choosing a location for the pump inlet is equally as critical as deciding on the location for a river grab sample. The design of the pump and pipework also requires careful design to avoid artefacts being introduced through the action of pumping the water. Dissolved oxygen concentration is difficult to sustain through a pumped system and GLC/MS facilities can detect micro-organic contaminants from the pipework and glands.

Passive Sampling

The use of passive samplers greatly reduces the cost and the need of infrastructure on the sampling location. Passive samplers are semi-disposable and can be produced at a relatively low cost, thus they can be employed in great numbers, allowing for a better cover and more data being collected. Due to being small the passive sampler can also be hidden, and thereby lower the risk of vandalism. Examples of passive sampling devices are the diffusive gradients in thin films (DGT) sampler, Chemcatcher, Polar organic chemical integrative sampler (POCIS), semipermeable membrane devices (SPMDs), stabilized liquid membrane devices (SLMDs), and an air sampling pump.

Remote Surveillance

Although on-site data collection using electronic measuring equipment is common-place, many monitoring programmes also use remote surveillance and remote access to data in real time. This requires the on-site monitoring equipment to be connected to a base station via either a telemetry network, land-line, cell phone network or other telemetry system such as Meteor burst. The advantage of remote surveillance is that many data feeds can come into a single base station for storing and analysis. It also enable trigger levels or alert levels to be set for individual monitoring sites and parameters so that immediate action can be initiated if a trigger level is exceeded. The use of remote surveillance also allows for the installation of very discrete monitoring equipment which can often be buried, camouflaged or tethered at depth in a lake or river with only a short whip aerial protruding. Use of such equipment tends to reduce vandalism and theft when monitoring in locations easily accessible by the public.

Remote Sensing

Environmental remote sensing uses aircraft or satellites to monitor the environment using multi-channel sensors.

There are two kinds of remote sensing. Passive sensors detect natural radiation that is emitted or reflected by the object or surrounding area being observed. Reflected sunlight is the most common source of radiation measured by passive sensors and in environmental remote sensing, the sensors used are tuned to specific wavelengths from far infrared through visible light frequencies to the far ultraviolet. The volumes of data that can be collected are very large and require dedicated computational support. The output of data analysis from remote sensing are false colour images which differentiate small differences in the radiation characteristics of the environment being monitored. With a skilful operator choosing specific channels it is possible to amplify differences which are imperceptible to the human eye. In particular it is possible to discriminate subtle changes in chlorophyll a and chlorophyll b concentrations in plants and show areas of an environment with slightly different nutrient regimes.

Active remote sensing emits energy and uses a passive sensor to detect and measure the radiation that is reflected or backscattered from the target. LIDAR is often used to acquire information about the topography of an area, especially when the area is large and manual surveying would be prohibitively expensive or difficult.

Remote sensing makes it possible to collect data on dangerous or inaccessible areas. Remote sensing applications include monitoring deforestation in areas such as the Amazon Basin, the effects of climate change on glaciers and Arctic and Antarctic regions, and depth sounding of coastal and ocean depths.

Orbital platforms collect and transmit data from different parts of the electromagnetic spectrum, which in conjunction with larger scale aerial or ground-based sensing and analysis, provides information to monitor trends such as El Niño and other natural long and short term phenomena. Other uses include different areas of the earth sciences such as natural resource management, land use planning and conservation.

Bio-monitoring

The use of living organisms as monitoring tools has many advantages. Organisms living in the environment under study are constantly exposed to the physical, biological and chemical influences of that environment. Organisms that have a tendency to accumulate chemical species can often accumulate significant quantities of material from very low concentrations in the environment. Mosses have been used by many investigators to monitor heavy metal concentrations because of their tendency to selectively adsorb heavy metals.

Similarly, eels have been used to study halogenated organic chemicals, as these are adsorbed into the fatty deposits within the eel.

Other Sampling Methods

Ecological sampling requires careful planning to be representative and as noninvasive as possible. For grasslands and other low growing habitats the use of a quadrat – a 1-metre square frame – is often used with the numbers and types of organisms growing within each quadrat area counted.

Sediments and soils require specialist sampling tools to ensure that the material recovered is representative. Such samplers are frequently designed to recover a specified volume of material and may also be designed to recover the sediment or soil living biota as well such as the Ekman grab sampler.

Data Interpretations

The interpretation of environmental data produced from a well designed monitoring programme is a large and complex topic addressed by many publications. Regrettably it is sometimes the case that scientists approach the analysis of results with a pre-conceived outcome in mind and use or misuse statistics to demonstrate that their own particular point of view is correct.

Statistics remains a tool that is equally easy to use or to misuse to demonstrate the lessons learnt from environmental monitoring.

Environmental Quality Indices

Since the start of science-based environmental monitoring, a number of quality indices have been devised to help classify and clarify the meaning of the considerable volumes of data involved. Stating that a river stretch is in "Class B" is likely to be much more informative than stating that this river stretch has a mean BOD of 4.2, a mean dissolved oxygen of 85%, etc. In the UK the Environment Agency formally employed a system called General Quality Assessment (GQA) which classified rivers into six quality letter bands from A to F based on chemical criteria and on biological criteria. The Environment Agency and its devolved partners in Wales (Countryside Council for Wales, CCW) and Scotland (Scottish Environmental Protection Agency, SEPA) now employ a system of biological, chemical and physical classification for rivers and lakes that corresponds with the EU Water Framework Directive.

Methods of Monitoring

A wide variety of methods are available for measuring contaminants in ambient air, with an equally wide variation in cost and precision. Specific monitoring methods should be chosen taking into consideration the purpose, objectives and budget of the monitoring programme.

Monitoring for the purposes of national environmental standards can only be carried out with the high-precision instrumental methods in accordance with Schedule 2 of the regulations. High-precision instrumental methods would also be generally used for research studies or other specific investigations, where there is a need to understand the ways in which contaminant levels fluctuate over short time periods (hours or days).

Ambient air quality guidelines carry recommended monitoring methods that should be used for the assessment of the contaminants covered by the guidelines.

There is still a place for monitoring methods that fall outside the specifications for standards and guideline assessments. Methods that involve lower resolution instruments can be used for an initial screening survey, where a detailed study is not warranted. These may be used as a prelude to more detailed monitoring. If contaminant levels are found to be low, then the same methods could be used for repeat surveys over subsequent years. Low resolution methods are also useful for prioritising a number of different areas that have already been selected for detailed monitoring programmes.

It should be stressed that screening methods cannot be used to determine compliance with national standards or guidelines. A screening method is any non-standard method that is used on an exploratory basis and provides indicative data for a particular contaminant. Despite this, the method still needs to have a level of accuracy and precision suitable for the purpose of monitoring (eg, Occupational Health and Safety equipment is not suitable for ambient air quality monitoring).

Screening methods are purely indicative methods and any results must be treated with caution. For example, a one-day-in-three monitoring regime of PM_{10} may indicate that there have been no breaches of the ambient air standard. However, this does not necessarily mean that exceedences were not occurring during the two days that were not sampled. A number of exceedences detected by this method would indicate detailed monitoring for national environmental standards is warranted.

Using Existing Monitoring for National Environmental Standards

Regional monitoring programmes conducted since the introduction of the RMA should, at least in the case of PM_{10}, provide a good indication of where national environmental standards are likely to be breached. However care must be taken when considering whether such ambient air quality monitoring can be used for the regulatory requirements of national environmental standards. The following points should be considered:

- Is the data capture sufficient and is there an appropriate time average?
- Are monitoring methods appropriate for standards?
- Are the appropriate contaminants being monitored?
- Is it at an appropriate location?

Earlier PM_{10} monitoring in New Zealand traditionally adopted a one-day-in-three or one-day-in-six approach. This approach is not suitable for national environmental standards because continuous monitoring of 24-hour averages from midnight to midnight is required. It may be found that some analysers that require manual filter changes are not practical for midnight changeovers, unless these analysers can have automated samplers retrofitted. Regional councils have shifted to continuous monitors since the NES for air quality came into effect.

It is not only necessary to choose analysers that are compliant with the relevant Australian/New Zealand standards or USEPA standard; it is also necessary to operate the analyser in accordance with that standard. Simply ensuring an analyser's specifications are compliant with the standard is not sufficient.

A large proportion of PM_{10} monitoring in New Zealand is conducted at residential neighbourhood sites. While these may not exactly align with the 'worst location' requirement in the NES for air quality, careful consideration should be made concerning relocation of the site. There may be little to gain by relocating a current site within the same general vicinity of an airshed if it will end a useful record of long-term data.

Choosing Appropriate Monitoring Equipment

Choosing the appropriate monitoring equipment is important for achieving the aims of the monitoring programme. Equipment that does not measure the contaminant in the required measurement range clearly will not provide useful data.

The following factors should be considered before purchasing monitoring equipment:

- Purpose of monitoring: Screening, compliance monitoring or research. Different sensitivities may apply to different types of monitors. For example, compliance monitoring may require a higher level of sensitivity and resolution than a monitor used for screening purposes.
- Duration of deployment: If the instrument is to be used for short-term screening surveys then portability, size, weight and robustness may be factors to consider.
- Detection limit, precision and measurement range: will the instrument be able to measure within the required range, based on the monitoring objective?
- Consumables: How frequently do parts need to be changed or replaced?
- Ability to conform to relevant monitoring standards: Does the instrument conform to monitoring standards such as those required in the NES for air quality?
- Frequency of calibration: The time taken to complete automatic zero and span checks, and operational and multipoint calibrations. Instruments that spend significant time on an automatic zero and span checks will reduce your percentage data recovery. An instrument that requires frequent calibration in order to comply with a particular standard will impose costs on an organisation (time and resources).
- Ease of use: Some instruments can be extremely difficult to operate, which imposes costs in terms of training and person-hours spent resolving problems.
- Communication: Ideally instruments should have ethernet/IP ports, analogue and RS232 connectivity to allow connection to dataloggers and telemetry.

- Ability to interface remotely: Most modern instruments will allow an operator to remotely operate an instrument.
- Environmental requirements: Does the instrument require air conditioning, temperature or humidity control?
- Cost: Consider both the cost of the instrument and the cost of consumables.
- Reliability: Find out who else has operated similar instruments and discuss the pros and cons with other organisations before committing to a purchase.
- Memory or on-board storage capacity: Determine if memory capacity is sufficient for your current and future needs (eg, memory expansion features).
- Instrument parameters that may be recorded in addition to output concentration: can the instrument record any other parameters (eg, temperature)?

Do your homework before making a purchase. For organisations that make decisions primarily based on price, consider imposing instrument specifications and standards as conditions of purchase.

Methodology Types

Monitoring methodologies can be divided into three categories according to cost and the level of accuracy and precision.

Continuous Monitoring Methods

These are high-resolution methods that provide continuous records of contaminant levels. They can operate over extended periods (weeks or months) with minimal operator intervention. Remote communication is possible by telemetry. They have a high degree of measurement precision, and have detection levels around one order of magnitude or more below typical background levels. As might be expected, these are the most expensive monitoring methods. A high standard of maintenance, calibration, and operational and quality control procedures are required for good data quality.

Gravimetric Particulate Methods

In the past, gravimetric particulate methods have formed the mainstay of particulate monitoring in New Zealand. The implementation of the NES for air quality, however, has shifted the use of these methods to the analysis of airborne lead, co-location studies and screening surveys.

Monitoring starts when a known volume of air is pumped through a pre-weighed filter for a known length of time (typically 24 hours). The filter is reweighed after exposure and a concentration determined. Most systems used in New Zealand require manual changes of the sampling filters between each sample, although a number of semi-automated systems are also available. This can be done on consecutive days. Manually changing filters at midnight, however, is operationally impractical when compliance monitoring.

Passive Monitoring Methods (Diffusion Tubes and Badges)

Diffusion tubes work when a contaminant is diffused into a tube containing either an adsorbent or reactive material. Analysis of the tubes following a known exposure time (typically two to four weeks) will provide a time-averaged contaminant concentration. Badges work in a similar way, the difference being the sampler configuration. Badges typically have higher uptake rates and are used more widely.

Because these methods are simple and cheap, they can provide a good picture of spatial variation over a large area. They are particularly useful in screening surveys and during the initial stages of an air quality monitoring programme.

Though a cheap screening tool, there are a number of limitations to this method, such as lower accuracy and no indication of peak levels. Quality control and assurance during laboratory analysis must be of the highest standard to attain consistent results. The results from passive samplers can be used in conjunction with high-resolution instruments to determine spatial variation across an airshed over a relevant averaging period. This method can also be useful for comparison with annual guidelines.

Historically, wet chemical methods were used to monitor levels of gaseous contaminants. These methods are no longer recommended due to interferences and the low resolution of data. High-resolution instrumental methods are now recommended.

Low-level instrumental methods using optical sensors are commonly used for monitoring occupational exposure. While such analysers may occasionally be useful for incident investigations, they should not be used for routine compliance monitoring due to their low sensitivity.

Mandatory Ambient Air Monitoring Methods

The following are mandatory monitoring methods under the NES for air quality. Note that these Australian/New Zealand and USEPA standards apply to monitoring methods. It is not sufficient to simply use an analyser that conforms to a standard: it is also necessary to operate the analyser in accordance with the operational requirements of that standard.

Carbon Monoxide (CO)

CO monitoring instruments are predominantly gas filter correlation infrared (GFC-IR) absorption analysers. This is the recommended CO monitoring method, although AS 3580.7.1 also allows for a non-dispersive infrared gas chromatograph with flame ionisation detector, or electrochemical sensor systems. These alternative methods, however, suffer from a variety of interfering species and are considered less robust than GFC-IR analysers.

In a GFC-IR analyser, ambient air is continuously sampled using a pump unit and the CO concentration in the sample air is measured by the absorption of infrared radiation at 4.5 to 4.9 nanometers (nm) wavelength. A reference detection system is used to alternately measure absorption due to CO in the ambient air stream and absorption by interfering species. An infrared detector and amplification system produce output voltages proportional to the CO concentration. The concentration is derived from the Beer–Lambert relation:

$$\mathbf{I_1 = I_0\, e^{-alc}}$$

where the sample passes through a cell tube of length 'l'. The analyser alternately measures the absorption I_0 of the air path with no CO present and the absorption I_1 of the ambient sample, with 'a' being the absorption coefficient, to provide the CO concentration, 'c'.

GFC instruments use a filter wheel to allow alternate measurement of total IR absorption. The analyser continually displays current CO concentrations, and, depending on the make and model of analyser, other parameters can be selected as necessary.

Environmental Sampling

Microbiologic Sampling of the Environment

Before 1970, U.S. hospitals conducted regularly scheduled culturing of the air and environmental surfaces (e.g., floors, walls, and table tops). By 1970, CDC and the American Hospital Association (AHA) were advocating the discontinuation of routine environmental culturing because rates of healthcare–associated infection had not been associated with levels of general microbial contamination of air or environmental surfaces, and because meaningful standards for permissible levels of microbial contamination of environmental surfaces or air did not exist. During 1970–1975, 25% of U.S. hospitals reduced the extent of such routine environmental culturing — a trend that has continued.

Random, undirected sampling (referred to as "routine" in previous guidelines) differs from the current practice of targeted sampling for defined purposes. Previous recommendations against routine sampling were not intended to discourage the use of sampling in which sample collection, culture, and interpretation are conducted in accordance with defined protocols. In this guideline, targeted microbiologic sampling connotes a monitoring process that includes:

- A written, defined, multidisciplinary protocol for sample collection and culturing.

- Analysis and interpretation of results using scientifically determined or anticipatory baseline values for comparison.
- Expected actions based on the results obtained.

Infection control, in conjunction with laboratorians, should assess the health-care facility's capability to conduct sampling and determine when expert consultation and services are needed.

Microbiologic sampling of air, water, and inanimate surfaces (i.e., environmental sampling) is an expensive and time-consuming process that is complicated by many variables in protocol, analysis, and interpretation. It is therefore indicated for only four situations. The first is to support an investigation of an outbreak of disease or infections when environmental reservoirs or fomites are implicated epidemiologically in disease transmission. It is important that such culturing be supported by epidemiologic data. Environmental sampling, as with all laboratory testing, should not be conducted if there is no plan for interpreting and acting on the results obtained. Linking microorganisms from environmental samples with clinical isolates by molecular epidemiology is crucial whenever it is possible to do so.

The second situation for which environmental sampling may be warranted is in research. Well-designed and controlled experimental methods and approaches can provide new information about the spread of health-care associated diseases. A classic example is the study of environmental microbial contamination that compared health-care associated infection rates in an old hospital and a new facility before and shortly after occupancy.

The third indication for sampling is to monitor a potentially hazardous environmental condition, confirm the presence of a hazardous chemical or biological agent, and validate the successful abatement of the hazard. This type of sampling can be used to:

- Detect bioaerosols released from the operation of health-care equipment (e.g., an ultrasonic cleaner) and determine the success of repairs in containing the hazard.
- Detect the release of an agent of bioterrorism in an indoor environmental setting and determine its successful removal or inactivation.
- Sample for industrial hygiene or safety purposes (e.g., monitoring a "sick building").

The fourth indication is for quality assurance to evaluate the effects of a change in infection-control practice or to ensure that equipment or systems perform according to specifications and expected outcomes. Any sampling for quality-assurance purposes must follow sound sampling protocols and address confounding factors through the use of properly selected controls. Results from a single environmental sample are difficult to interpret in the absence of a frame of reference or perspective. Evaluations of a change in infection-control practice are based on the assumption that the effect will be measured over a finite period, usually of short duration. Conducting quality-assurance sampling on an extended basis, especially in the absence of an adverse outcome, is usually unjustified. A possible exception might be the use of air sampling during major construction periods to qualitatively detect breaks in environmental infection-control measures. In one study, which began as part of an investigation of an outbreak of health-care associated aspergillosis, airborne concentrations of *Aspergillus* spores were measured in efforts to evaluate the effectiveness of sealing hospital doors and windows during a period of construction of a nearby building. Other examples of sampling for quality-assurance purposes may include commissioning newly constructed space in special care areas (i.e., ORs and units for immunosuppressed patients) or assessing a change in housekeeping practice. However, the only types of routine environmental microbiologic sampling recommended as part of a quality-assurance program are:

- The biological monitoring of sterilization processes by using bacterial spores.
- The monthly culturing of water used in hemodialysis applications and for the final dialysate use dilution.

Some experts also advocate periodic environmental sampling to evaluate the microbial/particulate quality for regular maintenance of the air handling system (e.g., filters) and to verify that the components of the system meet manufacturer's specifications. Certain equipment in health-care settings (e.g., biological safety cabinets) may also be monitored with air flow and particulate sampling to determine performance or as part of adherence to a certification program; results can then be compared with a predetermined standard of performance. These measurements, however, usually do not require microbiologic testing.

Air Sampling

Biological contaminants occur in the air as aerosols and may include bacteria, fungi, viruses, and pollens. Aerosols are characterized as solid or liquid particles suspended in air. Talking for 5 minutes and coughing each can produce 3,000 droplet nuclei; sneezing can generate approximately 40,000 droplets which then evaporate to particles in the size range of 0.5–12 μm. Particles in a biological aerosol usually vary in size from <1 μm to ≥50 μm. These particles may consist of a single, unattached organism or may occur in the form of clumps composed of a number of bacteria. Clumps can also include dust and dried organic or inorganic material. Vegetative forms of bacterial cells and viruses may be present in the air in a lesser number than bacterial spores or fungal spores. Factors that determine the survival of microorganisms within a bioaerosol include:

- The suspending medium.
- Temperature.

- Relative humidity.
- Oxygen sensitivity.
- Exposure to UV or electromagnetic radiation.

Many vegetative cells will not survive for lengthy periods of time in the air unless the protective cover (e.g., dried organic or inorganic matter). Pathogens that resist drying (e.g., *Staphylococcus* spp., *Streptococcus* spp., and fungal spores) can survive for long periods and can be carried considerable distances via air and still remain viable. They may also settle on surfaces and become airborne again as secondary aerosols during certain activities (e.g., sweeping and bed making).

Microbiologic air sampling is used as needed to determine the numbers and types of microorganisms, or particulates, in indoor air. Air sampling for quality control is, however, problematic because of lack of uniform air-quality standards. Although airborne spores of *Aspergillus* spp. can pose a risk for neutropenic patients, the critical number (i.e., action level) of these spores above which outbreaks of aspergillosis would be expected to occur has not been defined. Health-care professionals considering the use of air sampling should keep in mind that the results represent indoor air quality at singular points in time, and these may be affected by a variety of factors, including:

- Indoor traffic.
- Visitors entering the facility.
- Temperature.
- Time of day or year.
- Relative humidity.
- Relative concentration of particles or organisms, and g) the performance of the air-handling system components.

To be meaningful, air-sampling results must be compared with those obtained from other defined areas, conditions, or time periods.

Several preliminary concerns must be addressed when designing a microbiologic air sampling strategy. Because the amount of particulate material and bacteria retained in the respiratory system is largely dependent on the size of the inhaled particles, particle size should be determined when studying airborne microorganisms and their relation to respiratory infections. Particles >5 µm are efficiently trapped in the upper respiratory tract and are removed primarily by ciliary action. Particles ≤5 µm in diameter reach the lung, but the greatest retention in the alveoli is of particles 1–2 µm in diameter.

Bacteria, fungi, and particulates in air can be identified and quantified with the same methods and equipment. The basic methods include:

- Impingement in liquids.
- Impaction on solid surfaces.
- Sedimentation.
- Filtration.
- Centrifugation.
- Electrostatic precipitation.
- Thermal precipitation.

Of these, impingement in liquids, impaction on solid surfaces, and sedimentation (on settle plates) have been used for various air-sampling purposes in health-care settings.

Several instruments are available for sampling airborne bacteria and fungi. Some of the samplers are self-contained units requiring only a power supply and the appropriate collecting medium, but most require additional auxiliary equipment (e.g., a vacuum pump and an airflow measuring device [i.e., a flowmeter or anemometer]). Sedimentation or depositional methods use settle plates and therefore need no special instruments or equipment. Selection of an instrument for air sampling requires a clear understanding of the type of information desired and the particular determinations that must be made. Information may be needed regarding:

- One particular organism or all organisms that may be present in the air.
- The concentration of viable particles or of viable organisms.
- The change in concentration with time.
- The size distribution of the collected particles.

Before sampling begins, decisions should be made regarding whether the results are to be qualitative or quantitative. Comparing quantities of airborne microorganisms to those of outdoor air is also standard operating procedure. Infection-control professionals, hospital epidemiologists, industrial hygienists, and laboratory supervisors, as part of a multidisciplinary team, should discuss the potential need for microbial air sampling to determine if the capacity and expertise to conduct such sampling exists within the facility and when it is appropriate to enlist the services of an environmental microbiologist consultant.

Liquid impinger and solid impactor samplers are the most practical for sampling bacteria, particles, and fungal spores, because they can sample large volumes of air in relatively short periods of time.Solid impactor units are available as either "slit" or "sieve" designs. Slit impactors use a rotating disc as support for the collecting surface, which allows determinations of concentration over time. Sieve impactors commonly use stages with calibrated holes of different diameters. Some impactor-type samplers use centrifugal force to impact particles onto agar surfaces. The interior of either device must be made sterile to avoid inadvertent contamination from the sampler. Results obtained from either sampling device can be expressed as organisms or particles per unit volume of air (CFU/m^3).

Sampling for bacteria requires special attention, because bacteria may be present as individual organisms, as clumps, or mixed with or adhering to dust or covered with a protective coating of dried organic or inorganic substances. Reports of bacterial concentrations determined by air sampling therefore must indicate whether the results represent individual organisms or particles bearing multiple cells. Certain types of samplers (e.g., liquid impingers) will completely or partially disintegrate clumps and large particles; the sampling result will therefore reflect the total number of individual organisms present in the air.

The task of sizing a bioaerosol is simplified through the use of sieves or slit impactors because these samplers will separate the particles and microorganisms into size ranges as the sample is collected. These samplers must, however, be calibrated first by sampling aerosols under similar use conditions.

The use of settle plates (i.e., the sedimentation or depositional method) is not recommended when sampling air for fungal spores, because single spores can remain suspended in air indefinitely. Settle plates have been used mainly to sample for particulates and bacteria either in research studies or during epidemiologic investigations. Results of sedimentation sampling are typically expressed as numbers of viable particles or viable bacteria per unit area per the duration of sampling time (i.e., CFU/area/time); this method can not quantify the volume of air sampled. Because the survival of microorganisms during air sampling is inversely proportional to the velocity at which the air is taken into the sampler, one advantage of using a settle plate is its reliance on gravity to bring organisms and particles into contact with its surface, thus enhancing the potential for optimal survival of collected organisms. This process, however, takes several hours to complete and may be impractical for some situations.

Air samplers are designed to meet differing measurement requirements. Some samplers are better suited for one form of measurement than others. No one type of sampler and assay procedure can be used to collect and enumerate 100% of airborne organisms. The sampler and sampling method chosen should, however, have an adequate sampling rate to collect a sufficient number of particles in a reasonable time period so that a representative sample of air is obtained for biological analysis. Newer analytical techniques for assaying air samples include PCR methods and enzyme-linked immunosorbent assays (ELISAs).

Water Sampling

Water sampling in health-care settings is used detect waterborne pathogens of clinical significance or to determine the quality of finished water in a facility's distribution system. Routine testing of the water in a health-care facility is usually not indicated, but sampling in support of outbreak investigations can help determine appropriate infection-control measures.

Health-care facilities that conduct water sampling should have their samples assayed in a laboratory that uses established methods and quality-assurance protocols. Water specimens are not "static specimens" at ambient temperature; potential changes in both numbers and types of microbial populations can occur during transport. Consequently, water samples should be sent to the testing laboratory cold (i.e., at approximately 39.2 °F [4 °C]) and testing should be done as soon as practical after collection (preferably within 24 hours).

Because most water sampling in health-care facilities involves the testing of finished water from the facility's distribution system, a reducing agent (i.e., sodium thiosulfate [$Na_2S_2O_3$]) needs to be added to neutralize residual chlorine or other halogen in the collected sample. If the water contains elevated levels of heavy metals, then a chelating agent should be added to the specimen. The minimum volume of water to be collected should be sufficient to complete any and all assays indicated; 100 mL is considered a suitable minimum volume. Sterile collection equipment should always be used.

Sampling from a tap requires flushing of the water line before sample collection. If the tap is a mixing faucet, attachments (e.g., screens and aerators) must be removed, and hot and then cold water must be run through the tap before collecting the sample. If the cleanliness of the tap is questionable, disinfection with 500–600 ppm sodium hypochlorite (1:100 v/v dilution of chlorine bleach) and flushing the tap should precede sample collection.

Microorganisms in finished or treated water often are physically damaged ("stressed") to the point that growth is limited when assayed under standard conditions. Such situations lead to false-negative readings and misleading assessments of water quality. Appropriate neutralization of halogens and chelation of heavy metals are crucial to the recovery of these organisms. The choice of recovery media and incubation conditions will also affect the assay. Incubation temperatures should be closer to the ambient temperature of the water rather than at 98.6 °F (37 °C), and recovery media should be formulated to provide appropriate concentrations of nutrients to support organisms exhibiting less than rigorous growth. High-nutrient content media (e.g., blood agar and tryptic soy agar [TSA]) may actually inhibit the growth of these damaged organisms. Reduced nutrient media (e.g., diluted peptone and R2A) are preferable for recovery of these organisms.

Use of aerobic, heterotrophic plate counts allows both a qualitative and quantitative measurement for water quality. If bacterial counts in water are expected to be high in number (e.g., during waterborne outbreak investigations), assaying small quantities using pour plates or spread plates is appropriate. Membrane filtration is used when low-count specimens are expected and larger sampling volumes are required (≥100 mL). The sample is filtered through the membrane, and the filter is applied directly face-up onto the surface of the agar plate and incubated.

Unlike the testing of potable water supplies for coliforms (which uses standardized test and specimen collection parameters and conditions), water sampling to support epidemiologic investigations of disease outbreaks may be subjected to modifications dictated by the circumstances present in the facility. Assay methods for waterborne pathogens may also not be standardized. Therefore, control or comparison samples should be included in the experimental design. Any departure from a standard method should be fully documented and should be considered when interpreting results and developing strategies. Assay methods specific for clinically significant waterborne pathogens (e.g., *Legionella* spp., *Aeromonas* spp., *Pseudomonas* spp., and *Acinetobacter* spp.) are more complicated and costly compared with both methods used to detect coliforms and other standard indicators of water quality.

Environmental Surface Sampling

Routine environmental-surface sampling (e.g., surveillance cultures) in health-care settings is neither cost-effective nor warranted. When indicated, surface sampling should be conducted with multidisciplinary approval in adherence to carefully considered plans of action and policy.

Surface sampling is used currently for research, as part of an epidemiologic investigation, or as part of a comprehensive approach for specific quality assurance purposes. As a research tool, surface sampling has been used to determine:

- Potential environmental reservoirs of pathogens.
- Survival of microorganisms on surfaces.
- The sources of the environmental contamination.

Some or all of these approaches can also be used during outbreak investigations. Discussion of surface sampling of medical devices and instruments is beyond the scope of this document and is deferred to future guidelines on sterilization and disinfection issues.

Meaningful results depend on the selection of appropriate sampling and assay techniques. The media, reagents, and equipment required for surface sampling are available from any well-equipped microbiology laboratory and laboratory supplier. For quantitative assessment of surface organisms, non-selective, nutrient-rich agar media and broth (e.g., TSA and brain-heart infusion broth [BHI] with or without 5% sheep or rabbit blood supplement) are used for the recovery of aerobic bacteria. Broth media are used with membrane-filtration techniques. Further sample work-up may require the use of selective media for the isolation and enumeration of specific groups of microorganisms. Examples of selective media are MacConkey agar (MAC), Cetrimide agar, or Sabouraud dextrose- and malt extract agars and broths. Qualitative determinations of organisms from surfaces require only the use of selective or non-selective broth media.

Effective sampling of surfaces requires moisture, either already present on the surface to be sampled or via moistened swabs, sponges, wipes, agar surfaces, or membrane filters. Dilution fluids and rinse fluids include various buffers or general purpose broth media. If disinfectant residuals are expected on surfaces being sampled, specific neutralizer chemicals should be used in both the growth media and the dilution or rinse fluids. Lists of the neutralizers, the target disinfectant active ingredients, and the use concentrations have been published. Alternatively, instead of adding neutralizing chemicals to existing culture media (or if the chemical nature of the disinfectant residuals is unknown), the use of either:

- Commercially available media including a variety of specific and nonspecific neutralizers.
- Double-strength broth media will facilitate optimal recovery of microorganisms.

The inclusion of appropriate control specimens should be included to rule out both residual antimicrobial activity from surface disinfectants and potential toxicity caused by the presence of neutralizer chemicals carried over into the assay system.

Table: Examples of Eluents and Diluents for environmental-surface Sampling* +.

Solutions	Concentration in water
Ringer	1/4 strength
Peptone water	0.1%–1.0%
Buffered peptone water	0.067 M phosphate, 0.43% NaCl, 0.1% peptone
Phosphate-buffered saline	0.02 M phosphate, 0.9% NaCl
Sodium chloride (NaCl)	0.25%–0.9%
Calgon Ringer (This solution is used for dissolution of calcium alginate swabs.)	1/4 strength
Thiosulfate Ringer (This solution is used for neutralization of residual chlorine.)	1/4 strength
Water	n/a
Tryptic soy broth (TSB)	n/a
Brain-heart infusion broth (BHI) supplemented with 0.5% beef extract	n/a

+ A surfactant (e.g., polysorbate [i.e., Tween 80]) may be added to eluents and diluents. A concentration ranging from 0.01%–0.1% is generally used, depending on the specific application. Foaming may occur during use.

Several methods can be used for collecting environmental surface samples. Specific step-bystep discussions of each of the methods have been published. For best results, all methods should incorporate aseptic techniques, sterile equipment, and sterile recovery media.

Sample/rinse methods are frequently chosen because of their versatility. However, these sampling methods are the most prone to errors caused by manipulation of the swab, gauze pad, or sponge. Additionally, no microbiocidal or microbiostatic agents should be present in any of these items when used for sampling. Each of the rinse methods requires effective elution of microorganisms from the item used to sample the surface. Thorough mixing of the rinse fluids after elution (e.g., via manual or mechanical mixing using a vortex mixer, shaking with or without glass beads, and ultrasonic bath) will help to remove and suspend material from the sampling device and break up clumps of organisms for a more accurate count. In some instances, the item used to sample the surface (e.g., gauze pad and sponge) may be immersed in the rinse fluids in a sterile bag and subjected to stomaching. This technique, however, is suitable only for soft or absorbent items that will not puncture the bag during the elution process.

If sampling is conducted as part of an epidemiologic investigation of a disease outbreak, identification of isolates to species level is mandatory, and characterization beyond the species level is preferred. When interpreting the results of the sampling, the expected degree of microbial contamination associated with the various categories of surfaces in the Spaulding classification must be considered. Environmental surfaces should be visibly clean; recognized pathogens in numbers sufficient to result in secondary transfer to other animate or inanimate surfaces should be absent from the surface being sampled. Although the interpretation of a sample with positive microbial growth is self-evident, an environmental surface sample, especially that obtained from housekeeping surfaces, that shows no growth does not represent a "sterile" surface. Sensitivities of the sampling and assay methods (i.e., level of detection) must be taken into account when no-growth samples are encountered. Properly collected control samples will help rule out extraneous contamination of the surface sample.

Environmental Toxicity Testing

Toxicology is "the study of the adverse effects of chemical, physical, or biological agents on living organisms and the ecosystem." Most developed countries have enacted laws and regulations to control the marketing, labeling, and transportation of chemicals, pesticides, consumer products, medical products, food additives, and other substances of potential toxicological concern. Many of the provisions require manufacturers to conduct testing to identify potential hazards to human and animal health and the environment, and to submit the test data to regulatory authorities.

Government agencies conduct human health and ecological *risk assessments* to ascertain the effects of a chemical or other substance on human health and the environment, respectively. The processes involved in a risk assessment for human health can be broken down into four steps as illustrated in the diagram below from the US Environmental Protection Agency (EPA).

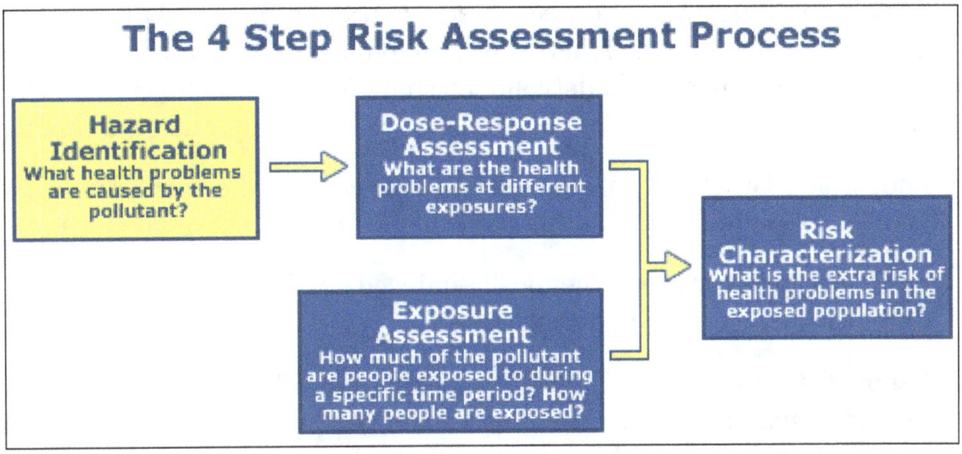

The hazard identification and dose-response assessment steps are primarily based on a number of different tests where animals are exposed to the chemical or test substance. These tests are called toxicity tests.

Government regulations often prescribe a specific regimen of toxicity testing to generate the data that enable regulators to determine the chemical's risks to human health and the environment. Typically, data for many different *toxicity endpoints* are needed for a regulatory submission on a new chemical or other regulated substance. Companies producing the chemical/product are responsible for the generation and submission of this "safety data" to regulatory authorities such as the US EPA, the European Chemicals Agency (ECHA), and Japan's Ministry of the Environment (MOE).

Toxicity Tests

A test method is a definitive procedure that produces a test result. A *toxicity test*, by extension, is designed to generate data concerning the adverse effects of a substance on human or animal health, or the environment. Many toxicity tests examine specific types of adverse effects, known as *endpoints*, such as eye irritation or cancer. Other tests are more general in nature, ranging from acute (single-exposure) studies to repeat dose (multiple-exposure) studies, in which animals are administered daily doses of a test substance.

Toxicity endpoints covered AltTox include the following:

- Acute Systemic Toxicity: Adverse effects occurring within a relatively short time after administration of a single (typically high) dose of a substance via one or more of the following exposure routes: oral, inhalation, skin, or injection.

- Carcinogenicity: Chemically-induced cancer, whether through genotoxic or non-genotoxic (e.g., growth-promoting) mechanisms.
- Dermal Penetration: Extent and rate by which a chemical is able to enter the body via the skin; also known as skin or percutaneous absorption.
- Ecotoxicity: Chemically-induced adverse effects on organisms in the environment, including mammals, birds, fish, amphibians, crustaceans, other aquatic invertebrates, and even plants; common study designs include acute systemic, dietary, and reproductive (also known as life-cycle) toxicity, and bioaccumulation.
- Endocrine Disruptors: Substances that interact with the hormonal systems of humans and wildlife, and thereby disrupt normal biological functions.
- Eye Irritation/Corrosion: Chemically-induced eye damage that is reversible (irritation) or irreversible (corrosion).
- Genotoxicity: Chemically-induced mutations and other alterations in the structure, information content, or segregation of genetic material (e.g., DNA strand breaks or a gain/loss in chromosome number).
- Neurotoxicity: Chemically-induced adverse effects on the brain, spinal cord, and peripheral nervous system (e.g., deficits in learning or sensory ability).

Pharmacokinetics and Metabolism

Study of the absorption, distribution, metabolism, and elimination (ADME) of drugs or chemicals in the body; also known as toxicokinetics.

Phototoxicity

Toxic responses from a substance (applied to the body or ingested) following exposure to light or skin irradiation.

Repeated Dose/Organ Toxicity

General toxicological effects occurring as a result of repeated daily exposure to a substance (via oral, inhalation, dermal, or injection routes) for a portion of the expected life span (i.e., subacute or subchronic exposure), or for the majority of the life span (i.e., chronic exposure).

Reproductive and Developmental Toxicity

Chemically-induced adverse effects on sexual function, fertility, and normal offspring development (e.g., spontaneous abortion, premature delivery, or birth defects); generally determined through the breeding of one or more generations of offspring.

Skin Irritation/Corrosion

Chemically-induced skin damage that is reversible (irritation) or irreversible (corrosion).

Skin Sensitization

The induction of allergic contact dermatitis following exposure to a chemical substance.

In addition to the traditional toxicity endpoints used for regulatory testing, toxicity test information for the following product categories are covered on AltTox.

Biologics, Vaccines and Medical Products

A variety of tests and endpoints related to efficacy, safety, and quality control testing of drugs, devices, biologics, diagnostics, and vaccines.

This category of testing is of considerable interest for the development of non-animal methods due to the large numbers of animals used.

Nanomaterials

A variety of endpoints and tests related to the safety and environmental impact of nanomaterials and products containing nanomaterials.

This category of testing is of considerable interest because of the uncertainty with the traditional animal toxicity test methods in providing hazard assessment data relevant to humans.

Cosmetics: Under development.

Green Chemistry

There is a long history in the use of animals as models for toxicity testing. During the 20th century, as regulatory agencies were established by national governments, animal test guidelines were developed to address regulatory requirements that products need to be "safe" for consumers. A review on pre-clinical testing by Parasuraman further explains the animal models and tests used for regulatory toxicity testing.

There are many advantages, including scientific, ethical, and economic ones, for replacing the animal toxicity tests with non-animal (in vitro and in silico) test systems. As science progresses, we have become aware of a number of problems with the use of animal models for assessing the safety of chemicals, drugs, and other products to human health.

Animal Welfare Considerations

Some conventional toxicity test methods consume hundreds to thousands of animals per substance examined. Statistics on animal use, such as the Seventh Report on the

Statistics on the Number of Animals used for Experimental and Other Scientific Purposes in the Member States of the European Union, indicate that toxicity testing accounts for a substantial portion of the more painful procedures experienced by animals (e.g., the use of death as the experimental endpoint in acute systemic toxicity studies).

Large government testing program requirements, such as REACH and HPV, have further increased the animal testing already conducted by companies. While REACH requires that animal testing is used "only when there are no other scientifically reliable ways of assessing the potential effects on humans or the environment," a report by the European Chemicals Agency in 2014 on the use of alternatives in testing for the REACH regulation explains that companies were not fully implementing the use of animal alternative methods, and that animal tests increased due to new registrations.

As public opposition towards animal testing has grown, some parts of the world have broadly prohibited testing on animals where alternative methods are "reasonably and practicably available". Animal testing bans may also be sector-specific, as in the case of the EU Cosmetics Regulation (EC), which now bans the marketing of any cosmetic product within the EU containing ingredients that have been tested on animals. Earlier EU test bans on cosmetics ingredients and final formulations had been imposed under the 7th Amendment to the EU Cosmetics Directive. Most recently, several additional countries have adopted bans on animal testing for cosmetic products, including India and Israel.

Time and Cost Considerations

Some conventional toxiciy tests take months to years to conduct and analyze (e.g., 4-5 years for carcinogenicity studies), and can cost thousands to millions of dollars per substance examined (e.g., $2-4 million per two-species carcinogenicity study).

In addition to the testing required for new chemicals and products, there is growing realization that the safety information we have on many existing chemicals is insufficient. While it is desirable to have only safe chemicals and products in the marketplace, the traditional means for conducting hazard assessments cannot keep pace with the demand.

Non-animal Toxicity Test Methods

The term "alternative," in the context of toxicity testing, has been used to describe new or revised test methods that result in the replacement of animal test methods, or that reduce the numbers of animals used, or refine the techniques to alleviate or minimize pain, distress, and suffering. Since the publication of this seminal work, governments, industry, NGOs, and other stakeholders have invested substantial time and financial resources to advance the 3Rs in research and testing.

At this time, the terms "non-animal" or "animal-free" are commonly used to describe Replacement methods. This avoids confusion due to the multiple uses of "alternatives."

Non-animal methods for producing toxicological data include the following: In vitro cell and tissue-based methods and models: Toxicity assays can be conducted using models developed with primary cells, cell lines, stem cells, 3-dimensional cultured cells, excised tissues, or cultured organs. Some cell-based methods have already achieved validation and international acceptance:

- In silico systems: Computer-based methods such as (quantitative) structure-activity relationship ((Q)SAR) models and read-across can be used to predict the biological/toxicological properties of a substance.
- Integrated testing *and* other emerging strategies: It is often necessary to use more than one non-animal test to assess a single biological endpoint; it is therefore necessary to develop strategies for optimally combining test methods to address specific information needs. Integrated testing strategies combine methods, such as in silicomethods and in vitro assays, along with appropriate statistical analysis, for the prediction of in vivo toxicity responses. An understanding of the mechanisms of toxicity and the cellular pathways involved are currently being investigated with the goal of obtaining more predictive integrated test strategies.

Other strategies for reducing animal testing requirements include:

- Human epidemiology or accidental exposures.
- Human volunteer studies.
- Existing toxicological information on a substance is recognized as sufficient for risk assessment purposes.
- Waiving of a requirement to conduct new testing because human exposure levels are below what is considered a significant risk to human health, or because testing would be difficult, impossible, or meaningless given the nature of the substance in question.

Scientific Validation, Regulatory Acceptance and International Harmonization

To facilitate the replacement of old tests with new ones, international authorities developed processes and criteria for evaluating new toxicity test methods to determine whether they can replace an existing method.

Key steps in the pathway from test method development to widespread use and acceptance of a new toxicity test method are the following – in their typical chronological order:

- Research and development consists of the basic and applied research involved in the genesis of a new test method. It might be undertaken and funded by biotech companies, contract testing laboratories, research institutes, regulated industry, or government bodies.

- Prevalidation is the process that aims to establish the mechanistic basis of a test, standardize and optimize the test protocol, evaluate within-lab variability, and define a model or procedure by which test results are used to predict the in vivo toxicological endpoint.

- Validation is the process to evaluate a defined set of criteria that includes a test method's transferability to other laboratories, between-lab variability and reproducibility, and reliability.

- Peer review or independent assessment is the formal evaluation process of the results from a validation study. Recommendations on the validity of the test method for the indicated purpose are made to the sponsoring validation authority.

- Regulatory acceptance processes differ region by region. In Europe, ESAC endorsement can lead to EU-wide acceptance, considering the legal requirement under Directive that non-animal alternatives be used preferentially. In the US, agencies individually decide whether to accept an ICCVAM recommendation. Most government agencies have the authority to accept whatever methods they find valid to meet their particular needs.

- International harmonization involves collaborative efforts through established organizations such as the OECD, ICH, and VICH to adopt the same testing requirements and test protocols, so that different countries and regions require and accept the same types of data for regulatory decisions. International acceptance in the form of an OECD Test Guideline is an important milestone for a validated non-animal test method, and can have a significant impact on reducing animal use in regulatory assessments.

Test method validation refers to the process of determining whether a toxicity test method (or test battery or test scheme) is reliable (reproducible) and relevant for its intended purpose. Relevance in this setting means "the extent to which the test method correctly measures or predicts the (biological) effect of interest."

Criteria and processes for toxicity test method validation were developed in the mid-1990's by validation authorities in the EU (European Centre for the Validation of Alternative Methods or ECVAM), the US (Interagency Coordinating Committee on the Validation of Alternative Methods or ICCVAM), and by the international Organisation for Economic Cooperation and Development (OECD).

Each validation authority, however, still has some distinct processes, and provides their own test method submission guidance:

- EURL ECVAM (EU).

- ICCVAM (US).

- JaCVAM (Japan).
- KoCVAM (Korea).

To optimize success, the requirements for regulatory acceptance should be considered during the design and conduct of a validation study. The validation authorities emphasize the importance of the involvement of regulatory authorities in all stages of the validation process.

According to the OECD, formal validation "contributes strongly to the international acceptance of any proposed test method," however, validation in not a requirement for the development of a method as a TG. The OECD recommends that the acceptance of any test method in the absence of validation should be accompanied by a written justification.

The OECD seeks to promote the "harmonization of international regulatory acceptance of adequately validated test methods". The OECD notes that regulatory acceptance practices differ by country and even among agencies within a particular country, so acceptance by one authority does not indicate universal acceptance. National regulatory authorities may accept a test that has not undergone formal validation, or they may not accept a test that has undergone formal validation. The OECD provides guidance on international regulatory acceptance for its member countries, and encourages worldwide mutual acceptance of data.

Formal validation as described in these documents may not be necessary for use of all test methods. All methods should be evaluated for reliability, reproducibility, sensitivity and specificity; however, the need for additional formal evaluation is context and use-dependent. In this light, OECD has developed a guidance document for describing non-standard test methods.

OECD, ECVAM, and ICCVAM also provide guidance on quality assurance and quality control issues relevant to the conduct of inter-laboratory validation studies, such as test substance coding/blindness, data audits, and the use of Good Laboratory Practice (GLP) (OECD, n.d.) and Good Cell Culture Practice (GCCP) guidelines.

Methods of Environmental Toxicity Testing

Biomarkers

The word *biomarker* is short for biological marker, a term that refers to a measurable event occurring in a biological system, such as the human body. This event is then interpreted as a reflection, or marker, of a more general state of the organism or of life expectancy. In occupational health, a biomarker is generally used as an indicator of health status or disease risk.

Biomarkers are used for in vitro as well as in vivo studies that may include humans. Usually, three specific types of biological markers are identified. Although a few biomarkers may be difficult to classify, usually they are separated into biomarkers of exposure, biomarkers of effect or biomarkers of susceptibility.

Table: Examples of biomarkers of exposure or biomarkers of effect that are used in toxicological studies in occupational health.

Sample	Measurement	Purpose
Exposure biomarkers		
Adipose tissue	Dioxin	Dioxin exposure
Blood	Lead	Lead exposure
Bone	Aluminium	Aluminium exposure
Exhaled breath	Toluene	Toluene exposure
Hair	Mercury	Methylmercury exposure
Serum	Benzene	Benzene exposure
Urine	Phenol	Benzene exposure
Effect biomarkers		
Blood	Carboxyhaemoglobin	Carbon monoxide exposure
Red blood cells	Zinc-protoporphyrin	Lead exposure
Serum	Cholinesterase	Organophosphate exposure
Urine	Microglobulins	Nephrotoxic exposure
White blood cells	DNA adducts	Mutagen exposure

Given an acceptable degree of validity, biomarkers may be employed for several purposes. On an individual basis, a biomarker may be used to support or refute a diagnosis of a particular type of poisoning or other chemically-induced adverse effect. In a healthy subject, a biomarker may also reflect individual hypersusceptibility to specific chemical exposures and may therefore serve as a basis for risk prediction and counselling. In groups of exposed workers, some exposure biomarkers can be applied to assess the extent of compliance with pollution abatement regulations or the effectiveness of preventive efforts in general.

Biomarkers of Exposure

An exposure biomarker may be an exogenous compound (or a metabolite) within the body, an interactive product between the compound (or metabolite) and an endogenous component, or another event related to the exposure. Most commonly, biomarkers of exposures to stable compounds, such as metals, comprise measurements of the metal concentrations in appropriate samples, such as blood, serum or urine. With volatile chemicals, their concentration in exhaled breath (after inhalation of contamination-free air) may be assessed. If the compound is metabolized in the body, one or more metabolites may be chosen as a biomarker of the exposure; metabolites are often determined in urine samples.

Modern methods of analysis may allow separation of isomers or congeners of organic compounds, and determination of the speciation of metal compounds or isotopic ratios of certain elements. Sophisticated analyses allow determination of changes in the structure of DNA or other macromolecules caused by binding with reactive chemicals.

Such advanced techniques will no doubt gain considerably in importance for applications in biomarker studies, and lower detection limits and better analytical validity are likely to make these biomarkers even more useful.

Particularly promising developments have occurred with biomarkers of exposure to mutagenic chemicals. These compounds are reactive and may form adducts with macromolecules, such as proteins or DNA. DNA adducts may be detected in white blood cells or tissue biopsies, and specific DNA fragments may be excreted in the urine. For example, exposure to ethylene oxide results in reactions with DNA bases, and, after excision of the damaged base, N-7-(2-hydroxyethyl)guanine will be eliminated in the urine. Some adducts may not refer directly to a particular exposure. For example, 8-hydroxy-2′-deoxyguanosine reflects oxidative damage to DNA, and this reaction may be triggered by several chemical compounds, most of which also induce lipid peroxidation.

Other macromolecules may also be changed by adduct formation or oxidation. Of special interest, such reactive compounds may generate haemoglobin adducts that can be determined as biomarkers of exposure to the compounds. The advantage is that ample amounts of haemoglobin can be obtained from a blood sample, and, given the four-month lifetime of red blood cells, the adducts formed with the amino acids of the protein will indicate the total exposure during this period.

Adducts may be determined by sensitive techniques such as high-performance lipid chromatography, and some immunological methods are also available. In general, the analytical methods are new, expensive and need further development and validation. Better sensitivity can be obtained by using the ^{32}P post labelling assay, which is a nonspecific indication that DNA damage has taken place. All of these techniques are potentially useful for biological monitoring and have been applied in a growing number of studies. However, simpler and more sensitive analytical methods are needed. Given the limited specificity of some methods at low-level exposures, tobacco smoking or other factors may impact significantly on the measurement results, thus causing difficulties in interpretation.

Exposure to mutagenic compounds, or to compounds which are metabolized into mutagens, may also be determined by assessing the mutagenicity of the urine from an exposed individual. The urine sample is incubated with a strain of bacteria in which a specific point mutation is expressed in a way that can be easily measured. If mutagenic chemicals are present in the urine sample, then an increased rate of mutations will occur in the bacteria.

Exposure biomarkers must be evaluated with regard to temporal variation in exposure and the relation to different compartments. Thus, the time frams represented by the biomarker, that is, the extent to which the biomarker measurement reflects past exposures and accumulated body burden, must be determined from toxicokinetic data in order to interpret the result. In particular, the degree to which the biomarker indicates retention in specific target organs should be considered. Although blood samples are often used for biomarker studies, peripheral blood is generally not regarded as a compartment as such, although it acts as a transport medium between compartments. The degree to which the concentration in the blood reflects levels in different organs varies widely between different chemicals, and usually also depends upon the length of the exposure as well as time since exposure.

Sometimes this type of evidence is used to classify a biomarker as an indicator of (total) absorbed dose or an indicator of effective dose (i.e., the amount that has reached the target tissue). For example, exposure to a particular solvent may be evaluated from data on the actual concentration of the solvent in the blood at a particular time following the exposure. This measurement will reflect the amount of the solvent that has been absorbed into the body. Some of the absorbed amount will be exhaled due to the vapour pressure of the solvent. While circulating in the blood, the solvent will interact with various components of the body, and it will eventually become subject to breakdown by enzymes. The outcome of the metabolic processes can be assessed by determining specific mercapturic acids produced by conjugation with glutathione. The cumulative excretion of mercapturic acids may better reflect the effective dose than will the blood concentration.

Life events, such as reproduction and senescence, may affect the distribution of a chemical. The distribution of chemicals within the body is significantly affected by pregnancy, and many chemicals may pass the placental barrier, thus causing exposure of the foetus. Lactation may result in excretion of lipid-soluble chemicals, thus leading to a decreased retention in the mother along with an increased uptake by the infant. During weight loss or development of osteoporosis, stored chemicals may be released, which can then result in a renewed and protracted "endogenous" exposure of target organs. Other factors may affect individual absorption, metabolism, retention and distribution of chemical compounds, and some biomarkers of susceptibility are available.

Biomarkers of Effect

A marker of effect may be an endogenous component, or a measure of the functional capacity, or some other indicator of the state or balance of the body or organ system, as affected by the exposure. Such effect markers are generally preclinical indicators of abnormalities.

These biomarkers may be specific or non-specific. The specific biomarkers are useful because they indicate a biological effect of a particular exposure, thus providing evidence that can potentially be used for preventive purposes. The non-specific biomarkers do not point to an individual cause of the effect, but they may reflect the total, integrated effect due to a mixed exposure. Both types of biomarkers may therefore be of considerable use in occupational health.

There is not a clear distinction between exposure biomarkers and effect biomarkers. For example, adduct formation could be said to reflect an effect rather than the exposure. However, effect biomarkers usually indicate changes in the functions of cells, tissues or the total body. Some researchers include gross changes, such as an increase in liver weight of exposed laboratory animals or decreased growth in children, as biomarkers of effect. For the purpose of occupational health, effect biomarkers should be restricted to those that indicate subclinical or reversible biochemical changes, such as inhibition of enzymes. The most frequently used effect biomarker is probably inhibition of cholinesterase caused by certain insecticides, that is, organophosphates and carbamates. In most cases, this effect is entirely reversible, and the enzyme inhibition reflects the total exposure to this particular group of insecticides.

Some exposures do not result in enzyme inhibition but rather in increased activity of an enzyme. This is the case with several enzymes that belong to the P450 family. They may be induced by exposures to certain solvents and polyaromatic hydrocarbons (PAHs). Since these enzymes are mainly expressed in tissues from which a biopsy may be difficult to obtain, the enzyme activity is determined indirectly in vivo by administering a compound that is metabolized by that particular enzyme, and then the breakdown product is measured in urine or plasma.

Other exposures may induce the synthesis of a protective protein in the body. The best example is probably metallothionein, which binds cadmium and promotes the excretion of this metal; cadmium exposure is one of the factors that result in increased expression of the metallothionein gene. Similar protective proteins may exist but have not yet been explored sufficiently to become accepted as biomarkers. Among the candidates for possible use as biomarkers are the so-called stress proteins, originally referred to as heat shock proteins. These proteins are generated by a range of different organisms in response to a variety of adverse exposures.

Oxidative damage may be assessed by determining the concentration of malondialdehyde in serum or the exhalation of ethane. Similarly, the urinary excretion of proteins with a small molecular weight, such as albumin, may be used as a biomarker of early kidney damage. Several parameters routinely used in clinical practice (for example, serum hormone or enzyme levels) may also be useful as biomarkers. However, many of these parameters may not be sufficiently sensitive to detect early impairment.

Another group of effect parameters relate to genotoxic effects (changes in the structure of chromosomes). Such effects may be detected by microscopy of white blood cells that undergo cell division. Serious damage to the chromosomes—chromosomal aberrations or formation of micronuclei—can be seen in a microscope. Damage may also be revealed by adding a dye to the cells during cell division. Exposure to a genotoxic agent can then be visualized as an increased exchange of the dye between the two chromatids of each chromosome (sister chromatid exchange). Chromosomal aberrations are related to an increased risk of developing cancer, but the significance of an increased rate of sister chromatid exchange is less clear.

More sophisticated assessment of genotoxicity is based on particular point mutations in somatic cells, that is, white blood cells or epithelial cells obtained from the oral mucosa. A mutation at a specific locus may make the cells capable of growing in a culture that contains a chemical that is otherwise toxic (such as 6-thioguanine). Alternatively, a specific gene product can be assessed (e.g., serum or tissue concentrations of oncoproteins encoded by particular oncogenes). Obviously, these mutations reflect the total genotoxic damage incurred and do not necessarily indicate anything about the causative exposure. These methods are not yet ready for practical use in occupational health, but rapid progress in this line of research would suggest that such methods will become available within a few years.

Biomarkers of Susceptibility

A marker of susceptibility, whether inherited or induced, is an indicator that the individual is particularly sensitive to the effect of a xenobiotic or to the effects of a group of such compounds. Most attention has been focused on genetic susceptibility, although other factors may be at least as important. Hypersusceptibility may be due to an inherited trait, the constitution of the individual, or environmental factors.

The ability to metabolize certain chemicals is variable and is genetically determined. Several relevant enzymes appear to be controlled by a single gene. For example, oxidation of foreign chemicals is mainly carried out be a family of enzymes belonging to the P450 family. Other enzymes make the metabolites more water soluble by conjugation (e.g., N-acetyltransferase and μ-glutathion-*S*-transferase). The activity of these enzymes is genetically controlled and varies considerably. As mentioned above, the activity can be determined by administering a small dose of a drug and then determining the amount of the metabolite in the urine. Some of the genes have now been characterized, and techniques are available to determine the genotype. Important studies suggest that a risk of developing certain cancer forms is related to the capability of metabolizing foreign compounds. Many questions still remain unanswered, thus at this time limiting the use of these potential susceptibility biomarkers in occupational health.

Other inherited traits, such as $alpha_1$-antitrypsin deficiency or glucose-6-phosphate dehydrogenase deficiency, also result in deficient defence mechanisms in the body, thereby causing hypersusceptibility to certain exposures.

Most research related to susceptibility has dealt with genetic predisposition. Other factors play a role as well and have been partly neglected. For example, individuals with a chronic disease may be more sensitive to an occupational exposure. Also, if a disease process or previous exposure to toxic chemicals has caused some subclinical organ damage, then the capacity to withstand a new toxic exposure is likely to be less. Biochemical indicators of organ function may in this case be used as susceptibility biomarkers. Perhaps the best example regarding hypersusceptibility relates to allergic responses. If an individual has become sensitized to a particular exposure, then specific antibodies can be detected in serum. Even if the individual has not become sensitized, other current or past exposures may add to the risk of developing an adverse effect related to an occupational exposure.

A major problem is to determine the joint effect of mixed exposures at work. In addition, personal habits and drug use may result in an increased susceptibility. For example, tobacco smoke usually contains a considerable amount of cadmium. Thus, with occupational exposure to cadmium, a heavy smoker who has accumulated substantial amounts of this metal in the body will be at increased risk of developing cadmium-related kidney disease.

Application in Occupational Health

Biomarkers are extremely useful in toxicological research, and many may be applicable in biological monitoring. Nonetheless, the limitations must also be recognized. Many biomarkers have so far been studied only in laboratory animals. Toxicokinetic patterns in other species may not necessarily reflect the situation in human beings, and extrapolation may require confirmatory studies in human volunteers. Also, account must be taken of individual variations due to genetic or constitutional factors.

In some cases, exposure biomarkers may not at all be feasible (e.g., for chemicals which are short-lived in vivo). Other chemicals may be stored in, or may affect, organs which cannot be accessed by routine procedures, such as the nervous system. The route of exposure may also affect the distribution pattern and therefore also the biomarker measurement and its interpretation. For example, direct exposure of the brain via the olfactory nerve is likely to escape detection by measurement of exposure biomarkers. As to effect biomarkers, many of them are not at all specific, and the change can be due to a variety of causes, including lifestyle factors. Perhaps in particular with the susceptibility biomarkers, interpretation must be very cautious at the moment, as many uncertainties remain about the overall health significance of individual genotypes.

In occupational health, the ideal biomarker should satisfy several requirements. First of all, sample collection and analysis must be simple and reliable. For optimal analytical quality, standardization is needed, but the specific requirements vary considerably. Major areas of concern include: preparation of the in- dividual, sampling procedure and sample handling, and measurement procedure; the latter encompasses technical factors, such as calibration and quality assurance procedures, and individual- related factors, such as education and training of operators.

For documentation of analytical validity and traceability, reference materials should be based on relevant matrices and with appropriate concentrations of toxic substances or relevant metabolites at appropriate levels. For biomarkers to be used for biological monitoring or for diagnostic purposes, the responsible laboratories must have well-documented analytical procedures with defined performance characteristics, and accessible records to allow verification of the results. At the same time, nonetheless, the economics of characterizing and using reference materials to supplement quality assurance procedures in general must be considered. Thus, the achievable quality of results, and the uses to which they are put, have to be balanced against the added costs of quality assurance, including reference materials, manpower and instrumentation.

Another requirement is that the biomarker should be specific, at least under the circumstances of the study, for a particular type of exposure, with a clear-cut relationship to the degree of exposure. Otherwise, the result of the biomarker measurement may be too difficult to interpret. For proper interpretation of the measurement result of an exposure biomarker, the diagnostic validity must be known (i.e., the translation of the biomarker value into the magnitude of possible health risks). In this area, metals serve as a paradigm for biomarker research. Recent research has demonstrated the complexity and subtlety of dose-response relationships, with considerable difficulty in identifying no-effect levels and therefore also in defining tolerable exposures. However, this kind of research has also illustrated the types of investigation and the refinement that are necessary to uncover the relevant information. For most organic compounds, quantitative associations between exposures and the corresponding adverse health effects are not yet available; in many cases, even the primary target organs are not known for sure. In addition, evaluation of toxicity data and biomarker concentrations is often complicated by exposure to mixtures of substances, rather than exposure to a single compound at the time.

Before the biomarker is applied for occupational health purposes, some additional considerations are necessary. First, the biomarker must reflect a subclinical and reversible change only. Second, given that the biomarker results can be interpreted with regard to health risks, then preventive efforts should be available and should be considered realistic in case the biomarker data suggests a need to reduce the exposure. Third, the practical use of the biomarker must be generally regarded as ethically acceptable.

Industrial hygiene measurements may be compared with applicable exposure limits. Likewise, results on exposure biomarkers or effect biomarkers may be compared to biological action limits, sometimes referred to as biological exposure indices. Such limits should be based on the best advice of clinicians and scientists from appropriate disciplines, and responsible administrators as "risk managers" should then take into account relevant ethical, social, cultural and economic factors. The scientific basis should, if possible, include dose-response relationships supplemented by information on variations in susceptibility within the population at risk. In some countries, workers and members of the general public are involved in the standard-setting process and provide important input, particularly when scientific uncertainty is considerable. One of the major uncertainties is how to define an adverse health effect that should be prevented—for example, whether adduct formation as an exposure biomarker by itself represents an adverse effect (i.e., effect biomarker) that should be prevented. Difficult questions are likely to arise when deciding whether it is ethically defensible, for the same compound, to have different limits for adventitious exposure, on the one hand, and occupational exposure, on the other.

The information generated by the use of biomarkers should generally be conveyed to the individuals examined within the physician-patient relationship. Ethical concerns must in particular be considered in connection with highly experimental biomarker analyses that cannot currently be interpreted in detail in terms of actual health risks. For the general population, for example, limited guidance exists at present with regard to interpretation of exposure biomarkers other than the blood-lead concentration. Also of importance is the confidence in the data generated (i.e., whether appropriate sampling has been done, and whether sound quality assurance procedures have been utilized in the laboratory involved). An additional area of special worry relates to individual hypersusceptibility. These issues must be taken into account when providing the feedback from the study.

All sectors of society affected by, or concerned with carrying out, a biomarker study need to be involved in the decision-making process on how to handle the information generated by the study. Specific procedures to prevent or overcome inevitable ethical conflicts should be developed within the legal and social frameworks of the region or country. However, each situation represents a different set of questions and pitfalls, and no single procedure for public involvement can be developed to cover all applications of exposure biomarkers.

Toxicology Testing using Animals

Toxicology testing is an important aspect of animal testing because it is used to ensure the safety of various substances such as drugs. These tests are conducted by pharmaceutical corporations and other establishments such as animal testing facilities who are hired to perform animal testing for private companies. Toxicology testing is an area of animal testing with approximately a million animals used per year in Europe alone. This type of testing comprises an estimated ten percent of all animal testing procedures that occur each year. The number of animals used for a single chemical is also quite large, which has prompted animal welfare groups to target this type of testing, among others. Roughly five thousand animals undergo testing for just a single chemical and the number is more than double that for pesticides.

Benefits of Toxicology Tests

Toxicology testing has many benefits, which include assessment of products such as:

- Pesticides.
- Food products.
- Additives.
- Chemicals.
- Pharmaceuticals.

The majority of toxicology testing comprises individual ingredients in a formulation rather than the final product. One of the challenges, however, is that manufacturers cite that such toxicology tests actually exaggerate the toxicity of a chemical or similar substance. Therefore, it is actually preferable for them to perform the toxicology testing again on the final product, which then allows them to receive a label indicating lower toxicity. Unfortunately, this means using additional animals to conduct testing on the finished product.

Criticisms of Toxicology Testing

One enormous criticism of these tests is that they are performed without anaesthetics, which means that animals can suffer a great deal. The reason for avoiding anaesthetic use is because the interaction of two or more drugs can affect the animal's metabolism and excretion of the drug. This means that it is more difficult to gauge the specific effects of the drug in question. The overestimation can be a poor indicator of toxicity, in part, because the doses tend to be quite high and are also performed on a relatively low number of animals in comparison with humans. Although thousands of animals tested for one chemical may initially sound like a large number, a product could be reaching millions of consumers. Manufacturers are essentially extrapolating results from a very small number of animals receiving high doses of a substance. Another criticism is that toxicology testing in animals isn't closely aligned enough with the human body for the results to be accurate in humans.

Practical aspects of Toxicology Testing

Animals may receive just one application of a substance or they may receive regular or repeated applications over the course of several months. For some animals, they receive the chemical for their entire lifespan. Applications vary but include:

- Orally dispensing the substance, using a tube or placing it in the animal's food.
- Application of the substance to the animal's skin.
- Dripping the substance into the animal's eyes.
- Injecting the substance intravenously or via the animal's muscles.
- Injecting the substance just under the skin.
- Forced inhalation of a substance through use of a mask or chamber.

Types of Toxicology Tests

There are many forms of toxicology tests used on animals. The Draize test has been a highly publicised one with regards to animal welfare groups, who have campaigned against the test. In the Draize test, a substance is dripped or similarly applied to an animal's eyes or skin. Rabbits are commonly used in Draize testing, where researchers will apply the substance and then assess any subsequent damage or irritation to the rabbit's eyes. Regulatory aspects include that the test must, however, be stopped if the animal appears to be in extreme suffering or distress. The procedure can also cause redness, bleeding and blindness in the animal. There are alternatives such as a version of the

Draize test that is thought to reduce suffering but as of yet, it has not fully replaced the Draize test.

The LD50 test is also known as the 'lethal dose test' or the 'fifty percent test.' Although it has since been replaced with other procedures that don't cause the same level of animal suffering or use the large number of animals, the test is still used frequently and abundantly around the world. This test examines the toxicity of a chemical or other substance by assessing the dosage needed to kill half of the animal test subjects.

Other toxicology tests include foods and food additives. These tests are extremely expensive, costing in the millions and taking several years to finish. Some of the tests last a mere month while others continue for several months or more. These tests, due to the nature of the products, are the most rigid. They assess overall toxicity to the body of the animal as well as any damage to eyes or skin. Other aspects include the potential to cause malignant growth or affect reproduction.

Protecting the Public

Toxicology testing remains mandatory in many countries to satisfy the requirements for public consumption, use or exposure of substances. Animal welfare supporters continue to condemn toxicology testing as cruel, painful and unnecessary. The research community disagrees and believes it is vital to safeguarding public health. Hopefully, the focus can be to reduce animal suffering as much as possible while still obtaining the benefits of toxicology testing.

Biological Monitoring

Biological monitoring (biomonitoring) in occupational safety and health is the detection of substances (biomarkers) in biological samples of workers, compared to reference values. Biomonitoring can help in exposure assessment of specific chemicals, characterisation of exposure pathways and potential risks. Biomarkers can detect the exposure, the effect, or reveal susceptibility. Biomonitoring may be interpreted at group or individual level. Most common media are urine and blood. While multitudes of substances can be measured, there are still only limited numbers of validated methods and limit values with scientifically proven background. The first paper on occupational biomonitoring was published in the USA and worldwide mainstreaming started in the 1980s.

Biomonitoring is the repeated measurement of certain exposure-related chemical/biochemical markers in biological samples of subjects. Biomarkers are early (reversible) signs of exposure, effect or susceptibility with possible adverse health outcome. Biomonitoring is aimed at health protection, exposure and risk assessment. Characteristics of biomonitoring and workplace air monitoring (also known as environmental monitoring) are summarised in table.

Table: Comparison of biological and workplace air monitorings.

	Biological Monitoring	Workplace Air Monitoring
Quantifying	Internal dose	External dose.
Absorption	All routes	Inhalation only.
Confounders	Metabolic phenotype	Personal protective equipment, substances with similar structure/chemical properties.
Standardisation	Difficult	Easy.
Interpretation	Difficult	Moderately difficult.
Measurement	Indirect (biomarkers)	Usually indirect (dangerous substance).
Ethical issues	Important	None.
Variability	High	Usually low.

Classification of Biomarkers

Biomarkers are classified into three categories depending on their use or the specific context in which the test is being used.

A biomarker of exposure is the substance, or its metabolite, or the product of an interaction that is measured in a compartment or a bodily fluid. For example lead in blood may fairly represent the recent lead exposure of the individual.

A biomarker of effect is a measurable alteration (biochemical, structural, functional, behavioral, etc.) in an organism that can be associated with an established or potential health impairment or disease. Biomarkers of early disease indicate early biochemical or functional alterations, ranging from natural adaptation to disease. For example the value of Zinc protoporphyrin in blood is increased when lead exposure caused changes in the production of haemoglobin.

Biomarkers of genotoxicity (chromosomal aberrations, micronuclei, Comet test) are used to measure exposure to genotoxic chemicals, usually at group level. They are sensitive but not specific indicators and generally inadequate for occupational risk assessment purposes.

A biomarker of susceptibility is the marker of an ability to adversely respond to the challenge of exposure to a chemical. Genes can make certain individuals more vulnerable to toxins such as lead. An example for a susceptibility biomarker is the type of genetic code for 6-aminolevulinic acid dehydratase (ALAD), an enzyme involved in the toxicity of lead that exists in two forms. These effect-modifying factors can be inherent or acquired. Biomarkers of susceptibility are not generally used in routine biomonitoring.

Exposure and Biotransformation

The three main exposure pathways are inhalation (lungs), dermal (skin) and gastrointestinal (ingestion). Biological monitoring considers the overall systemic exposure (internal dose) and effect (biological effective dose) regardless of sources. Biomarkers may also reflect circumstances like change in atmospheric pressure, co-exposures and respiratory rate (e.g. heavy workload), which may all lead to higher (or decreased) uptake of the substance.

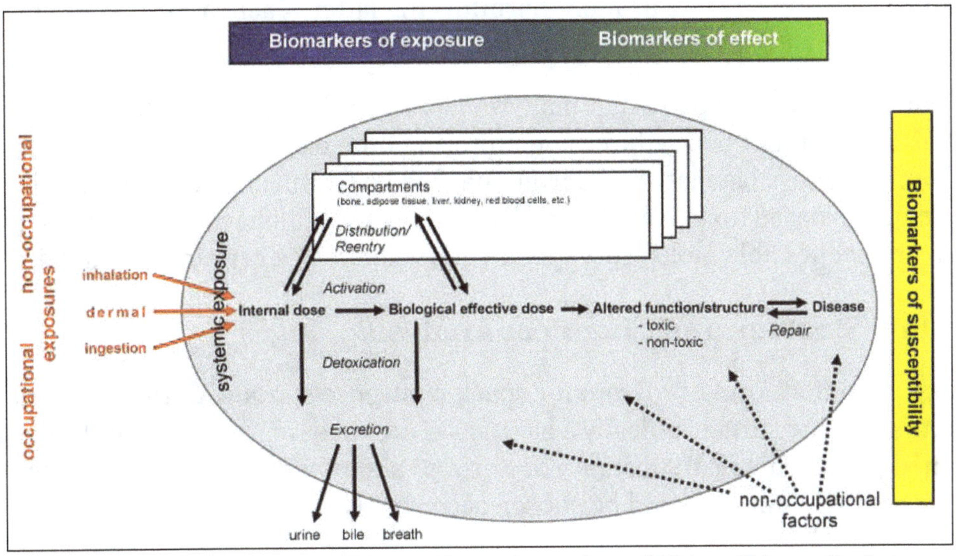

The process of exposure and biotransformation and possibilities of biomonitoring.

Once it enters the body the chemical (and its metabolites) can be:

- Distributed among body compartments.
- Undergo various modifications (biotransformation).
- Cause functional changes and diseases.
- Get excreted or deposited.

All these pathways are particular to the chemical substance concerned and may be specific to the individual. They may be influenced by internal or external non-occupational confounders.

Biomarkers can be used effectively if their toxicological background is understood:

- The fate of the chemical and its metabolites in the body (toxicokinetics).
- The mechanism of the disease/adverse effect (toxicodynamics).
- The way the individual factor promotes the chemical to cause disease/adverse effect (susceptibility).

In the process of biotransformation the external chemical substances are transformed in the body (metabolism). Enzyme activities are the fundamentals of susceptibility, including those involved in absorption and excretion. Individual varieties in the features of enzymes may result in different speeds and pathways in the metabolism of substances, modifying the dose-response curve. There are many modifiers in biotransformation: sex, age, body mass, co-exposures (workplace, non-occupational: e.g. diet, including fat, alcohol, medication, etc.). Compared to the original substance, its degradation products may be less or more harmful (the latter = activation). These metabolites may be used as biomarkers (of exposure or effect) as well.

Toxicokinetic features explain that many substances (although identifiable in the metabolic process) cannot be used for biomonitoring due to short half-life and disappearance from the accessible matrix, e.g. from the blood or the urine. Mathematical (such as physiologically-based toxicokinetic) models simulate metabolic processes in the human body to predict possible biological exposure parameters in a certain exposure scenario.

Sampling Strategy and Chemical Analysis

Planning a biomonitoring program for specific purpose starts with the selection of an adequate sampling strategy. Ideally a biomarker of exposure is specific for the exposure concerned (single chemical substance or a group) and reliably detectable, using non-invasive sampling. Timing should be chosen accordingly.

Periodical examinations are set either at fixed intervals or according to the previous biomonitoring results.

The analytical method is critical for the validity/reliability of a biomarker: accuracy, precision, reproducibility, recovery, sensitivity and specificity all have high influence to the consistency with the limit and reference values concerned. Several factors may affect the quality of the samples and the measurement of biomarkers: type of the matrix, point in time of collection, containers and preservatives and other additives used to stabilise the sample, storage temperature and transport time. Adequate reference material should be applied in adequate concentration range and matrix. Reliable and validated analytical methods must be used supported by internal quality control (IQC) and external quality assurance schemes (EQA).

The success of a biomonitoring program highly depends on good cooperation of partners: workers, the employer and the engineer, the occupational physician/industrial hygienist, the laboratory.

Biomarker Half-lives and Timing of Sampling

In the different matrices biomarkers have different half-lives, which is the time it takes the concentration to drop to half. Even a single biomarker may have different half-lives on different time scales: e.g. a quick concentration drop within days, followed by a slow decrease in years. Half-lives of biomarkers and the feasibility of single-compartment or multi-compartment model determine appropriate sampling strategy: The timing and the frequency of sample collection. For substances with long half-life, during continuous exposure, concentration reaches a certain value that remains stable (equilibrium) and reflects long-term intake. In such a case the timing of sample collection during a working day or even working week is not important. However, enough time must be left to reach equilibrium (e.g. lead, cadmium).

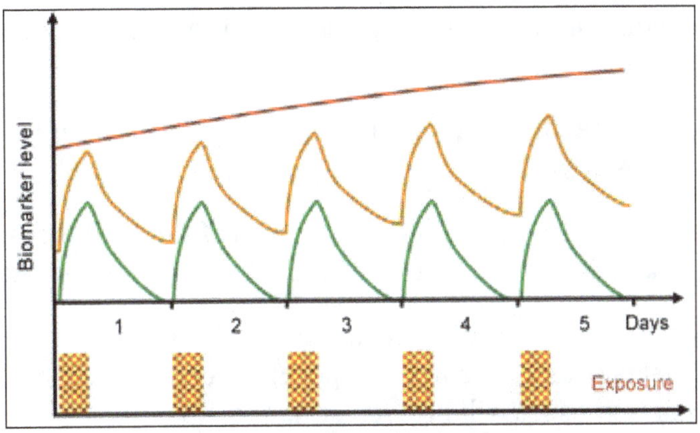

Different half-lives of biomarkers.

When the half-life is short (a few hours – a day) the concentration varies remarkably during the working day or week, thus timing of sampling is essential. Such a concentration reflects exposure over a short time and may not be representative of average long-term exposure. A meaningful picture of the average exposure may be obtained only from several samples.

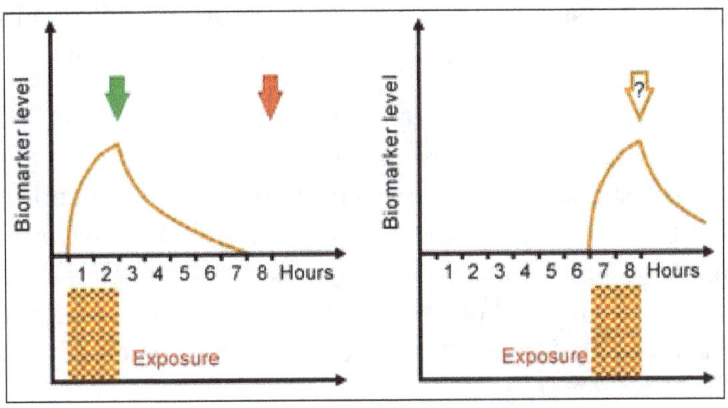

Possible pitfalls in biomonitoring of irregular exposures and short-half lives of biomarkers.

Usually sampling takes place:

- At the end of shift for short half-live biomarkers.
- At the end of the work-week for short half-live biomarkers with tendency to cumulate.
- Any time for highly cumulating biomarkers with long half-lives.

Samples can be also taken prior to shift or during shift in certain circumstances.

As a general rule, sampling is to be undertaken at a time when the inner and the external exposures are in a state of equilibrium which is not to be expected in cases where exposures/activities are short (maintenance, etc.). In such cases, sampling shall be performed at the end of the relevant activity. In case of substances with short half-lives (<4 hours), validity of biomonitoring highly depends on the actual exposure profile during the shift.

Next figure illustrates the challenge of biomonitoring for different exposures to substances with short half-lives:

- If exposure takes place in the first two hours of the shift, subsequent sampling assesses peak exposure, while end of shift sampling misses detection.
- If exposure takes place in the last two hours of the shift, sampling at the end of the shift detects peak exposure. If exposure is erroneously reckoned as all-day, the total daily exposure will be overestimated.

Biological Matrices

Biomonitoring is available in many kinds of biological media (matrices): Urine, blood, exhaled air, saliva, sweat, semen, faeces, and several tissues. The appropriate matrix depends on the type of biomarker (exposure, effect, susceptibility) and the type of chemical (parent compound or metabolite, volatile or non-volatile, hydrophilic or hydrophobic, unstable or persistent, etc.). The most important biological samples used in the everyday practice are urine, blood, and occasionally exhaled air.

urine collection is more readily accepted by workers (easy and not invasive). Concentration in urine usually reflects the mean plasma level of the substance since last urination. Rapidly excreted substances (e.g. solvents) are detected in end-of-shift samples. Timed collected specimens may be more representative. In routine urinary biomonitoring spot samples adjusted for dilution (by creatinine or relative density) are adequate alternatives. Highly diluted or very concentrated urine (creatinine: <0.3, >3.0g/litre; relative density: <1.010, >1.030) are not suitable for such adjustment, thus new specimens should be collected.

Most absorbed substance and active metabolites can be found in blood, which is the second most common biological matrix used in routine biomonitoring. Unlike urine, composition of blood is regulated within narrow limits (e.g. pH, concentrations of natural substances), thus it seldom needs adjustment. Blood sampling is useful for inorganic chemicals (e.g. metals) and for organic chemicals that are poorly metabolised and have a sufficiently long half-life.

Exhaled air analysis is non-invasive and most appropriate to estimate exposure to volatile organic substances (e.g. solvents), although it is much less frequently used.

Hair has become an important biological sample in forensic, environmental and clinical toxicology (reconstructing past exposures), its use in routine occupational biomonitoring is not advisable due high contamination risk.

Setting of Biological Limit Values

Different approaches of scientific bodies resulted in various value definitions, also determining their application: BLVs, BATs and BEIs are set by the above mentioned scientific bodies, and are equally used in European countries. However, the recommended limit values are not necessarily mainstreamed into national legislations. The diversity is huge in Member States' policies, applications, values and measures. The sole specific mandatory EU-wide policy on biomonitoring is on lead.

The International Commission on Occupational Health (ICOH) defined Biological Limit Value as the biomarker level that can be directly associated with (the lack of) a biological effect or disease. The European Scientific Committee on Occupational Exposure Limits (SCOEL) defined: 'A Biological Limit Value (BLV) is a reference value for the evaluation of potential health risk in the practice of occupational health.Exposure concentrations equivalent to the BLV generally do not affect the health of the employee adversely, when they are attained regularly under workplace conditions (8 hrs/day, 5 days/week), except in cases of hypersensitivity.' 'It is presented as the concentration in the appropriate biological medium of the relevant agent, its metabolite, or indicator of effect.

Health-based Biological Limit Values (BLVs) are obtained in biological effect monitoring or linking no-observed-adverse effect-level to adverse effects human studies. Although these values are preferred, the number of such biomarkers is limited. Many SCOEL BLVs are obtained from the corresponding Occupational Exposure Limits (OELs) by matching the 'mean' level of a biological index with the corresponding OEL (concentration limit in the workplace air). These values are calculated from studies comparing exposures (OEL) and the corresponding biological concentrations observed. For non-traditional working schedules (not 8 hours/day, 5 days/week) BLVs can be derived from toxicokinetic and toxicodynamic bases. When an OEL serves as protection against non-systemic effects (e.g. respiratory irritation), and also for substances with significant non-inhalatory exposure routes, the BLV is set to avoid systemic effects (e.g. intoxication), and is not derived from the OEL.

BAT (biological tolerance) values are based either on the relationship between external exposure and internal dose, or between the internal dose and the resulting effect of the substance, using the average of internal dose. BLVs and BAT can be used at group level and, with certain restrictions, at the individual level too.

Traditionally, carcinogenic substances are excluded from setting BAT levels as no safe level can be set. However, a BLV could theoretically be derived for carcinogens with an OEL or based on correlation of workplace air concentration and body burden (e.g.

Germany, Expositionsäquivalente für krebserzeugende Arbeitsstoffe EKA: Exposure equivalents for carcinogenic substances). Carcinogens may have a BLW value.

Differences between BLV and BAT (Biologischer Arbeitsstoff-Toleranzwert or Biological Tolerance Value by the Deutsche Forschungsgemeinschaft – DFG, Germany) disappeared following the harmonisation of definitions. Initially BATs used to be maximum permissible values.

In certain circumstances proper BLV cannot be established because of missing data. In this case setting of limit value for intervention for substances is based on occupational-medical information together with toxicological data. However, it does not exclude a risk of adverse health effects (e.g. BLW – Biologischer Leit-Wert by the Deutsche Forschungsgemeinschaft – DFG, Germany). E.g. BLW may refer to the non-carcinogenic effect of an otherwise carcinogenic chemical.

Biological Exposure Indices of the American Conference of Governmental Industrial Hygienists (ACGIH) are indirect guidance values based mainly on the correlation between the biomarker concentration and the airborne concentration of the original substance.

The benchmark dose approach uses a mathematical model including all the dose-response data.

In the USA, most BEIs are also derived from the corresponding OELs the basis of mean values. These results can be interpreted only at the group level: some individuals may exceed the value due to variables other than exposure. Intervention is required only if BEI values are exceeded for a longer period of time or by a substantial group of exposed.

Variability of Results

Only accurate and reproducible analytical results should be used for biomonitoring and health surveillance. The predictive value of an effect biomarker is the extent to which that particular biomarker is capable of correctly separating subjects with a likelihood of impairment or disease from those without it and it may be influenced by different factors.

Biomarker concentration levels decrease when moving on from exposure to disease: this increases measurement uncertainty. Measuring the actual chemical can limit biological variability, but the advantage is that the measurement is not confounded by intermediary biological steps or by identical metabolites of other reactions. However, measuring the metabolites may be better if the parent chemical is unstable or volatile, or the metabolite is directly involved in the mechanism of toxicity (needs activation).

There may be a substantial variation among values obtained from different workers on the same day and also from day to day for a single worker: this may make dose-response relationships harder to assess.

Biological limit values in occupational medicine are set for healthy adults. There is no general rule for setting biological limit values for workers with certain health impairments. The disease may influence the biomarker level directly (e.g. biomarkers of effect), or the disease may alter the uptake, metabolism or excretion of the chemical. For these workers biomonitoring needs to be reconsidered and an individual approach may be necessary.

Certain vulnerable groups may require stricter limit values: e.g. blood lead in fertile women and young workers.

Reference Values

Biomarker concentration values in the general (occupationally not exposed, but otherwise identical) population are useful reference intervals in the assessment of thresholds concerning occupational exposure. Although distinguishing between occupational and environmental exposure is very difficult (if possible at all), these reference values can help identifying subjects with an increased exposure.

The International Union of Pure and Applied Chemistry (IUPAC) recommends the highest concentration (usually 95th percentile) of the biomarker in the unexposed population as reference limits (same as biological guidance values as defined by SCOEL and *Biologische Arbeitsstoff-Referenzwerte* of DFG). Reference values are influenced by: local environmental exposure levels, genetic polymorphisms, biological conditions (sex, age, fatty mass and diseases), and metabolic interferences due to lifestyle (tobacco and alcohol consumption, diet, medications). Thus it is recommended that reference values should be defined at the local level (country, region). Due to improvement in chemical analysis methods, a decreasing trend in occupational exposure levels and the change in concentrations of environmental pollutants (including air, water, food, soil, etc.), reference values are continuously being adapted.

Use of Biological Monitoring

Only highly specific, validated biomonitoring tools should be used as part of routine medical surveillance programs, especially when individual decisions have to be made. Selection of appropriate biomarkers depends on the targeted aims and possibilities. The benefit to workers is highest if the biomarker is close to the target organ and the test has high predictive value.

Biological monitoring plays a key role in occupational risk assessment, complementing workplace air monitoring. It can be used as a part of medical health surveillance. Biomonitoring can be used to perform or validate risk assessment when other approaches are unavailable or inadequate.

Biological monitoring may assess individual exposure to chemicals by different routes, providing additional information for occupational risk assessment from hazard identification to the control of prevention measures. It can be used to identify exposed workers, providing strong indirect evidence for the presence of the chemical in the workplace. Several factors contribute to the disruption of expected correlations between biomonitoring and workplace air monitoring, but biomonitoring, not workplace air monitoring, actually reflects systemic exposure. Biomonitoring may better assess the degree of recent and, also past exposure to chemicals from all routes and clarify dose-response relationships. Use at compounds with a skin notation may benefit the most, because it allows monitoring of dermal uptake of the substance.

Biomonitoring can help in the correct interpretation of doubtful clinical tests if the dose-response relationship is known. However, this has been controversially discussed.

There may be varied consequences of biomonitoring results above the reference value. Depending on the exposure, biomarker concerned and Member State policies it may result in:

- A more frequent biomonitoring.
- Detailed health surveillance of the affected or all exposed workers.
- Revision of the risk assessment.
- Removal of the worker from the exposure/job.

Furthermore, biological monitoring data can be used very fruitfully when setting DNEL/DMEL (Derived No-effect Level/ Derived Minimal Effect Level) and preparing the exposure scenario according to REACH. If a reliable dose-response relationship exists between biomarker concentration and effects in human, the 'No Observed Adverse Effect Level' (NOAEL) expressed as the biomarker level may directly serve as a basis for the DNEL.

Interpretation of Biomonitoring Results

Biomarker concentration is compared to appropriate reference values, but the user should be aware of all the uncertainty of the results and how the reference is defined. A biomarker above the reference value does not mean disease or hazard but potentially higher exposure than the reference population. This, however, may not necessarily mean increased health risk, but only if confirmed with repeated measurements and analysis of circumstances (SCOEL). Collective biomarker data (e.g. BEI) should be used only for the assessment of exposure at group level while individual data (e.g. BAT) are appropriate for drawing conclusions on personal exposure.

Dose-response relationships of effects can be compared to the cumulated frequency distribution of the biomarker. The probability to observe a certain effect (blue, orange or red) is achieved by projecting 100% of the cumulated frequency distribution to intersect the dose-response curves: 80% for blue effect, 35% for orange effect, and 0% for red effect. Using an individual result, the probabilities can be calculated for a single worker.

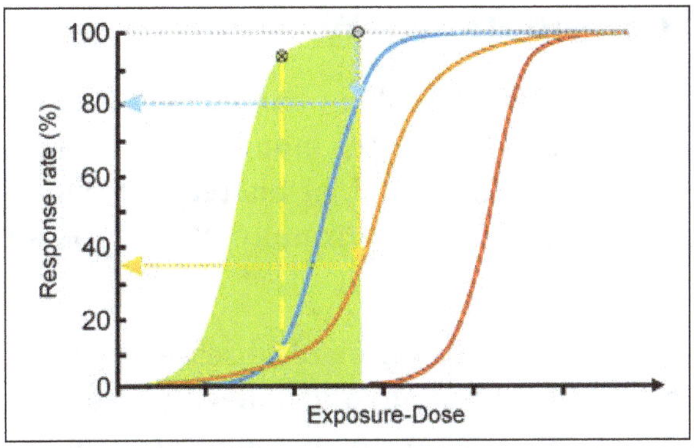

Interpretation of biomonitoring results at the group and at the individual levels.

Interpretation can be easier if variability factors are taken into consideration: uncovering and explaining variability may become a resource more than a limitation and may provide useful information for the interpretation of the results. Special attention should be paid to the thorough explanation of the use of biomarkers of carcinogenicity, emphasising that they are neither biomarkers nor direct predictors of cancer disease.

Ethical Considerations

Individual biomonitoring results are medical data, thus should be handled accordingly. Interpretation of the results is the task of a physician trained in the field of occupational health and informed on the entire situation (workplace exposures and worker's health). Data handling, storage and communication are specified in Directive 98/24/EC and in further European and national legislations. Communication (and interpretation) of individual results should be made only towards the worker concerned. Group data can be communicated towards the employer and workers' representatives. The ethical considerations must be regarded during the entire process of a biomonitoring study.

ICOH's code of ethics states that 'Biomarkers must be chosen for their validity and relevance for protection of the health of the worker concerned, with due regard to their sensitivity, their specificity and their predictive value'. Principles of this internationally agreed document are:

- Biomonitoring should not be used as screening tests or for insurance purposes'.
- Current knowledge in biomonitoring of susceptibility does not justify job opportunity discrimination of affected workers.
- Priority should be given to non-invasive (urine) and easily collected sampling (spot). In these cases informed consent is usually not required for routine procedures using validated biomarkers.
- Invasive tests or tests posing a health risk calls for a risk-benefit analysis, and informed consent of the worker.

Information and Consultation of Workers

According to Directive 98/24/EC, workers have a privileged right to be informed and know of the results of any biological monitoring in the framework of health surveillance. The individual worker shall, at his request, have access to the health and exposure records relating to him personally. Where, as a result of health surveillance:

- A worker is found to have an identifiable disease or adverse health effect which is considered by a doctor or occupational health-care professional to be the result of exposure at work to a hazardous chemical agent.
- A binding biological limit value is found to have been exceeded.

The worker shall be informed by the doctor or other suitably qualified person of the result which relates to him personally, including information and advice regarding any health surveillance which he should undergo following the end of the exposure.

Stricter requirements apply, especially as regards record-keeping and information of workers, when carcinogens or mutagens' exposure occur, laid out in Directive 2004/36/EC (carcinogens and mutagens at work). For example, all cases of occupational cancers shall be notified to the competent authority. Records shall be kept for at least 40 years following the end of exposure,

Advantages and Disadvantages of Biological Monitoring

'Without biomonitoring as a tool occupational risk assessment of chemicals would be more uncertain and vague than it is'.

A unique advantage of biomonitoring as compared to air monitoring is that it takes into consideration:

- All exposure routes (including inhalation, ingestion and dermal absorption) and circumstances (physical activity, multiple and uneven exposures).
- The individual response (variability in absorption/metabolism/excretion).

Biomonitoring is also an adequate tool in the assessment of the efficacy of protective equipment, ventilation and other hygiene measures. Biomonitoring may help estimating past exposure and assessing individual susceptibility. Specific and sensitive biomarkers may enable intervention in due time, thus preventing the development of a disease.

The above mentioned advantages also have some drawbacks. Biomarkers:

- Are usually unable to specify the source of the exposure (occupational or non-occupational).

- May not be sufficiently specific to a particular chemical.
- Are not suitable for identification of workplace contaminations in general.
- May be interfered by other chemicals in the biological medium (e.g. Medications).
- Are not useful at all for the assessment/monitoring of acute and local toxic effects (e.g. Irritation).
- And the provision of samples for biomonitoring may be a burden for workers (e.g. blood samples).
- The integrated use of both workplace air and biological monitoring is considered still to be the best approach to individual exposure assessment.

Bioremediation

Bioremediation is a process used to treat contaminated media, including water, soil and subsurface material, by altering environmental conditions to stimulate growth of microorganisms and degrade the target pollutants. In many cases, bioremediation is less expensive and more sustainable than other remediation alternatives. Biological treatment is a similar approach used to treat wastes including wastewater, industrial waste and solid waste.

Most bioremediation processes involve oxidation-reduction reactions where either an electron acceptor (commonly oxygen) is added to stimulate oxidation of a reduced pollutant (e.g. hydrocarbons) or an electron donor (commonly an organic substrate) is added to reduce oxidized pollutants (nitrate, perchlorate, oxidized metals, chlorinated solvents, explosives and propellants). In both these approaches, additional nutrients, vitamins, minerals, and pH buffers may be added to optimize conditions for the microorganisms. In some cases, specialized microbial cultures are added (bioaugmentation) to further enhance biodegradation. Some examples of bioremediation related technologies are phytoremediation, mycoremediation, bioventing, bioleaching, landfarming, bioreactor, composting, bioaugmentation, rhizofiltration, and biostimulation.

Chemistry

Most bioremediation processes involve oxidation-reduction (Redox) reactions where a chemical species donates an electron (electron donor) to a different species that accepts the electron (electron acceptor). During this process, the electron donor is said to be oxidized while the electron acceptor is reduced. Common electron acceptors in bioremediation processes include oxygen, nitrate, manganese (III and IV), iron (III), sulfate, carbon dioxide and some pollutants (chlorinated solvents, explosives, oxidized metals, and radionuclides). Electron donors include sugars, fats, alcohols, natural organic material, fuel hydrocarbons and a variety of reduced organic pollutants. The redox potential for common biotransformation reactions is shown in the table.

Process	Reaction	Redox potential (E_h in mV)
Aerobic	$O_2 + 4e^- + 4H^+ \rightarrow 2H_2O$	600 ~ 400
Anaerobic		
Denitrification	$2NO_3^- + 10e^- + 12H^+ \rightarrow N_2 + 6H_2O$	500 ~ 200
Manganese Iv Reduction	$MnO_2 + 2e^- + 4H^+ \rightarrow Mn^{2+} + 2H_2O$	400 ~ 200
Iron Iii Reduction	$Fe(OH)_3 + e^- + 3H^+ \rightarrow Fe^{2+} + 3H_2O$	300 ~ 100
Sulfate Reduction	$SO_4^{2-} + 8e^- + 10 H^+ \rightarrow H_2S + 4H_2O$	0 ~ −150
Fermentation	$2CH_2O \rightarrow CO_2 + CH_4$	−150 ~ −220

Aerobic

Aerobic bioremediation is the most common form of oxidative bioremediation process where oxygen is provided as the electron acceptor for oxidation of petroleum, polyaromatic hydrocarbons (PAHs), phenols, and other reduced pollutants. Oxygen is generally the preferred electron acceptor because of the higher energy yield and because oxygen is required for some enzyme systems to initiate the degradation process. Numerous laboratory and field studies have shown that microorganisms can degrade a wide variety of hydrocarbons, including components of gasoline, kerosene, diesel, and jet fuel. Under ideal conditions, the biodegradation rates of the low- to moderate-weight aliphatic, alicyclic, and aromatic compounds can be very high. As the molecular weight of the compound increases, so does the resistance to biodegradation.

Common approaches for providing oxygen above the water table include landfarming, composting and bioventing. During landfarming, contaminated soils, sediments, or sludges are incorporated into the soil surface and periodically turned over (tilled) using conventional agricultural equipment to aerate the mixture. Composting accelerates pollutant biodegradation by mixing the waste to be treated with a bulking agent, forming into piles, and periodically mixed to increase oxygen transfer. Bioventing is a process that increases the oxygen or air flow into the unsaturated zone of the soil which increases the rate of natural in situ degradation of the targeted hydrocarbon contaminant.

Approaches for oxygen addition below the water table include recirculating aerated water through the treatment zone, addition of pure oxygen or peroxides, and air sparging. Recirculation systems typically consist of a combination of injection wells or galleries and one or more recovery wells where the extracted groundwater is treated, oxygenated, amended with nutrients and reinjected. However, the amount of oxygen that can be provided by this method is limited by the low solubility of oxygen in water (8 to 10 mg/L for water in equilibrium with air at typical temperatures). Greater amounts of oxygen can be provided by contacting the water with pure oxygen or addition of hydrogen peroxide (H_2O_2) to the water. In some cases, slurries of solid calcium or magnesium peroxide are injected under pressure through soil borings. These solid peroxides react with water releasing H_2O_2 which then decomposes releasing oxygen. Air sparging involves the injection of air under pressure below the water table. The air injection pressure must be great enough to overcome the hydrostatic pressure of the water and resistance to air flow through the soil.

Anaerobic

Anaerobic bioremediation can be employed to treat a broad range of oxidized contaminants including chlorinated ethenes (PCE, TCE, DCE, VC), chlorinated ethanes (TCA, DCA), chloromethanes (CT, CF), chlorinated cyclic hydrocarbons, various energetics (e.g., perchlorate, RDX, TNT), and nitrate. This process involves the addition of an electron donor to: 1) Deplete background electron acceptors including oxygen, nitrate, oxidized iron and manganese and sulfate; and 2) Stimulate the biological and chemical reduction of the oxidized pollutants. Hexavalent chromium (Cr[VI]) and uranium (U[VI]) can be reduced to less mobile and less toxic forms (e.g., Cr[III], U[IV]). Similarly, reduction of sulfate to sulfide (sulfidogenesis) can be used to precipitate certain metals (e.g., zinc, cadmium). The choice of substrate and the method of injection depend on the contaminant type and distribution in the aquifer, hydrogeology, and remediation objectives. Substrate can be added using conventional well installations, by direct-push technology, or by excavation and backfill such as permeable reactive barriers (PRB) or biowalls. Slow-release products composed of edible oils or solid substrates tend to stay in place for an extended treatment period. Soluble substrates or soluble fermentation products of slow-release substrates can potentially migrate via advection and diffusion, providing broader but shorter-lived treatment zones. The added organic substrates are first fermented to hydrogen (H_2) and volatile fatty acids (VFAs). The VFAs, including acetate, lactate, propionate and butyrate, provide carbon and energy for bacterial metabolism.

Heavy Metals

Heavy metals including cadmium, chromium, lead and uranium are elements so they cannot be biodegraded. However, bioremediation processes can potentially be used to reduce the mobility of these material in the subsurface, reducing the potential for human and environmental exposure. The mobility of certain metals including chromium (Cr) and uranium (U) varies depending on the oxidation state of the material. Microorganisms can be used to reduce the toxicity and mobility of chromium by reducing hexavalent chromium, Cr(VI) to trivalent Cr (III). Uranium can be reduced from the more mobile U(VI) oxidation state to the less mobile U(IV) oxidation state. Microorganisms are used in this process because the reduction rate of these metals is often slow unless catalyzed by microbial interactions Research is also underway to develop methods to remove metals from water by enhancing the sorption of the metal to cell walls. This approach has been evaluated for treatment of cadmium, chromium, and lead. Phytoextraction processes concentrate contaminants in the biomass for subsequent removal.

Additives

In the event of biostimulation, adding nutrients that are limited to make the environment more suitable for bioremediation, nutrients such as nitrogen, phosphorus, oxygen, and carbon may be added to the system to improve effectiveness of the treatment.

Many biological processes are sensitive to pH and function most efficiently in near neutral conditions. Low pH can interfere with pH homeostasis or increase the solubility of toxic metals. Microorganisms can expend cellular energy to maintain homeostasis or cytoplasmic conditions may change in response to external changes in pH. Some anaerobes have adapted to low pH conditions through alterations in carbon and electron flow, cellular morphology, membrane structure, and protein synthesis.

Limitations

Bioremediation can be used to completely mineralize organic pollutants, to partially transform the pollutants, or alter their mobility. Heavy metals and radionuclides are elements that cannot be biodegraded, but can be bio-transformed to less mobile forms. In some cases, microbes do not fully mineralize the pollutant, potentially producing a more toxic compound. For example, under anaerobic conditions, the reductive dehalogenation of TCE may produce dichloroethylene (DCE) and vinyl chloride (VC), which are suspected or known carcinogens. However, the microorganism *Dehalococcoides* can further reduce DCE and VC to the non-toxic product ethene. Additional research is required to develop methods to ensure that the products from biodegradation are less persistent and less toxic than the original contaminant. Thus, the metabolic and chemical pathways of the microorganisms of interest must be known. In addition, knowing these pathways will help develop new technologies that can deal with sites that have uneven distributions of a mixture of contaminants.

Also, for biodegradation to occur, there must be a microbial population with the metabolic capacity to degrade the pollutant, an environment with the right growing conditions for the microbes, and the right amount of nutrients and contaminants. The biological processes used by these microbes are highly specific, therefore, many environmental factors must be taken into account and regulated as well. Thus, bioremediation processes must be specifically made in accordance to the conditions at the contaminated site. Also, because many factors are interdependent, small-scale tests are usually performed before carrying out the procedure at the contaminated site. However, it can be difficult to extrapolate the results from the small-scale test studies into big field operations. In many cases, bioremediation takes more time than other alternatives such as land filling and incineration.

Phytoremediation

Phytoremediation refers to the technologies that use living plants to clean up soil, air, and water contaminated with hazardous contaminants. It is defined as "the use of green plants and the associated microorganisms, along with proper soil amendments and agronomic techniques to either contain, remove or render toxic environmental contaminants harmless".

Phytoremediation is proposed as a cost-effective plant-based approach of remediation that takes advantage of the ability of plants to concentrate elements and compounds from the environment and to detoxify various compounds. The concentrating effect results from the ability of certain plants called hyperaccumulators to bioaccumulate chemicals. The remediation effect is quite different. Toxic heavy metals cannot be degraded, but organic pollutants can be and are generally the major targets for phytoremediation. Several field trials confirmed the feasibility of using plants for environmental cleanup.

Application

Phytoremediation may be applied to polluted soil or static water environment. Examples where phytoremediation has been used successfully include the restoration of abandoned metal mine workings, and sites where polychlorinated biphenyls have been dumped during manufacture and mitigation of ongoing coal mine discharges reducing the impact of contaminants in soils, water, or air. Contaminants such as metals, pesticides, solvents, explosives, and crude oil and its derivatives, have been mitigated in phytoremediation projects worldwide. Many plants such as mustard plants, alpine pennycress, hemp, and pigweed have proven to be successful at hyperaccumulating contaminants at toxic waste sites.

Not all plants are able to accumulate heavy metals or organics pollutants due to differences in the physiology of the plant. Even cultivars within the same species have varying abilities to accumulate pollutants.

This technology has been increasingly investigated and has been employed at sites with soils contaminated with lead, uranium, and arsenic. While it has the advantage that environmental concerns may be treated in situ, one major disadvantage of phytoremediation is that it requires a long-term commitment, as the process is dependent on a plant's ability to grow and thrive in an environment that is not ideal for normal plant growth.

Advantages and Limitations

- Advantages:
    - In principle, plants that engage in phytoremediation of toxic elements could be harvested, thus removing these elements from the polluted site.

  - The cost of the phytoremediation is lower than that of traditional processes both *in situ* and *ex sit*.
  - The plants can be easily monitored.
  - The possibility of the recovery and re-use of valuable metals (by companies specializing in "phyto mining").
  - It is potentially the least harmful method because it uses naturally occurring organisms and preserves the environment in a more natural state.
  - It preserves the topsoil, maintaining the fertility of the soil.
  - Increase soil health, yield, and plant phytochemicals.
  - The use of plants also reduces erosion and metal leaching in the soil.
- Limitations:
  - Phytoremediation simply relocates toxic heavy metals, it does not remove them from the locale.
  - For heavy metals, phytoremediation simply relocates the toxic element, it does not remove them.
  - Phytoremediation is limited to the surface area and depth occupied by the roots.
  - Slow growth and low biomass require a long-term commitment.
  - With plant-based systems of remediation, it is not possible to completely prevent the leaching of contaminants into the groundwater (without the complete removal of the contaminated ground, which in itself does not resolve the problem of contamination).
  - The survival of the plants is affected by the toxicity of the contaminated land and the general condition of the soil.
  - Bio-accumulation of contaminants, especially metals, into the plants which then pass into the food chain, from primary level consumers upwards or requires the safe disposal of the affected plant material.
  - When taking up heavy metals, sometimes the metal is bound to the soil organic matter, which makes it unavailable for the plant to extract

Processes

A range of processes mediated by plants or algae are tested in treating environmental problems.

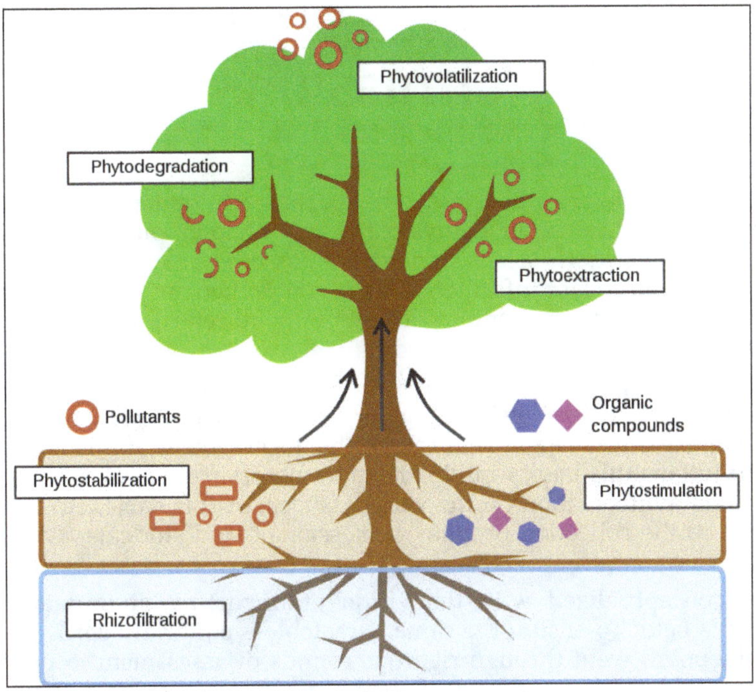

Phytoremediation process.

References

- Good-practice-guide-air-quality-monitoring-and-data-management, air: mfe.govt.nz, Retrieved 16 April, 2020
- Sampling, infectioncontrol-guidelines-environmental-background: cdc.gov, Retrieved 18 June, 2020
- Toxicity-testing-overview: alttox.org, Retrieved 31 March, 2020
- Toxicology-test-methods, toxicology-57951: iloencyclopaedia.org, Retrieved 19 February, 2020
- Toxicology-testing-using-animals: aboutanimaltesting.co.uk, Retrieved 13 January, 2020
- Overview-of-the-standard-methods-for-soil-ecotoxicology-testing, invertebrates-experimen-tal-models-in-toxicity-screening: intechopen.com, Retrieved 29 May, 2020

Permissions

All chapters in this book are published with permission under the Creative Commons Attribution Share Alike License or equivalent. Every chapter published in this book has been scrutinized by our experts. Their significance has been extensively debated. The topics covered herein carry significant information for a comprehensive understanding. They may even be implemented as practical applications or may be referred to as a beginning point for further studies.

We would like to thank the editorial team for lending their expertise to make the book truly unique. They have played a crucial role in the development of this book. Without their invaluable contributions this book wouldn't have been possible. They have made vital efforts to compile up to date information on the varied aspects of this subject to make this book a valuable addition to the collection of many professionals and students.

This book was conceptualized with the vision of imparting up-to-date and integrated information in this field. To ensure the same, a matchless editorial board was set up. Every individual on the board went through rigorous rounds of assessment to prove their worth. After which they invested a large part of their time researching and compiling the most relevant data for our readers.

The editorial board has been involved in producing this book since its inception. They have spent rigorous hours researching and exploring the diverse topics which have resulted in the successful publishing of this book. They have passed on their knowledge of decades through this book. To expedite this challenging task, the publisher supported the team at every step. A small team of assistant editors was also appointed to further simplify the editing procedure and attain best results for the readers.

Apart from the editorial board, the designing team has also invested a significant amount of their time in understanding the subject and creating the most relevant covers. They scrutinized every image to scout for the most suitable representation of the subject and create an appropriate cover for the book.

The publishing team has been an ardent support to the editorial, designing and production team. Their endless efforts to recruit the best for this project, has resulted in the accomplishment of this book. They are a veteran in the field of academics and their pool of knowledge is as vast as their experience in printing. Their expertise and guidance has proved useful at every step. Their uncompromising quality standards have made this book an exceptional effort. Their encouragement from time to time has been an inspiration for everyone.

The publisher and the editorial board hope that this book will prove to be a valuable piece of knowledge for students, practitioners and scholars across the globe.

Index

www.ingramcontent.com/pod-product-compliance
Lightning Source LLC
LaVergne TN
LVHW080629210926
848485LV00003B/236

9781639895298